T0345002

Selling Power

Markets and Governments in Economic History

A SERIES EDITED BY PRICE FISHBACK

Also in the series:

Selling Power

Economics, Policy, and
Electric Utilities Before 1940

JOHN L. NEUFELD

THE UNIVERSITY OF CHICAGO PRESS   CHICAGO AND LONDON

The University of Chicago Press, Chicago 60637
The University of Chicago Press, Ltd., London
© 2016 by The University of Chicago
All rights reserved. Published 2016.
Printed in the United States of America

25 24 23 22 21 20 19 18 17 16   1 2 3 4 5

ISBN-13: 978-0-226-39963-8 (cloth)
ISBN-13: 978-0-226-39977-5 (e-book)
DOI: 10.7208/chicago/9780226399775.001.0001

Library of Congress Cataloging-in-Publication Data

Names: Neufeld, John L., author.
Title: Selling power : economics, policy, and electric utilities before 1940 /
    John L. Neufeld.
Other titles: Markets and governments in economic history.
Description: Chicago ; London : The University of Chicago Press, 2016. |
Series: Markets and governments in economic history | Includes bibliographical
    references and index.
Identifiers: LCCN 2016015753 | ISBN 9780226399638 (cloth : alk. paper) |
    ISBN 9780226399775 (e-book)
Subjects: LCSH: Electric utilities—United States—History.
Classification: LCC HD9685.U5 N45 2016 | DDC 333.793/2097309041—dc23 LC record
    available at https://lccn.loc.gov/2016015753

To my family, Karol, Paul, Adam, and Margot, and to all the men and women who contributed to building America's electric utility industry

Contents

Illustrations

Tables

Acknowledgments

I am grateful to Will Hausman, Stephen Holland, Carl Kitchens, Al Link, Ken Snowden, and two anonymous reviewers who provided helpful comments, suggestions, and criticisms on drafts of this work. Any remaining errors are, of course, my own. I thank Bruno Sternegg for allowing me to use his photograph of a factory illustrating the use of shafts and belts. For many years, I benefited from a fruitful collaboration with my friend Will Hausman on projects that contributed to this book. I know the book would have benefitted from his full participation in its development, but, alas, his interests took him elsewhere. I am especially indebted to my wife, Karol. Not only has she put up with me for many years, but she also read the entire manuscript and made numerous suggestions. Her skill as a poet and as the best grammarian I know helped clarify my sometimes-obtuse writing, and I could not survive without her love and support.

The Economics of Electric Utilities

When Thomas Edison was born in 1847, the Civil War was more than a decade in the future. He was seventeen when the conclusion of that war freed millions of Americans from slavery. At age thirty-two, Edison publicly demonstrated his invention of an electric incandescent light bulb. The quality and versatility of that source of light led to its replacing the single source of artificial light on which humankind had depended since prehistory: the open flame. Other designs for a workable incandescent light bulb emerged about the same time as Edison's, and at least one can credibly claim to have preceded his. Edison's real invention was a technological system in which a utility centrally generated electricity and distributed it to customers for incandescent lighting. His incandescent bulb was the first designed to work in that system, and all of his competitors quickly altered their designs to be like his. Edison may not deserve credit for the invention of the incandescent bulb, but he does deserve credit for something much larger: he invented the modern electric utility industry, an industry that profoundly changed the way the world lives and works.

Thomas Edison died in 1931, three days before the fifty-second anniversary of the demonstration of his electric light. President Hoover suggested the nation mourn together the loss of the Great Inventor by forgoing the use of electricity for one minute. It was already apparent that this exercise had to be voluntary because, as the president recognized, the loss of electricity for even one minute could have been a threat to human life.[1] Within the span of the life of a single remarkable individual, the electric utility industry had moved from a gleam in his eye to a necessity so fundamental that, like air, even a brief interruption of access to its product was a serious threat.

The development of the electric power industry involved the struggles that accompanied huge technological changes. It includes many stories of individuals and the challenges they faced in bringing electricity to all Americans. Electric utilities profoundly changed the nation's economy and the ways Americans lived and worked. Although all of these are interesting and deserve a full treatment, no single book can achieve this. Although this book will touch upon these issues, its focus is on the structure of the industry's economic institutions and the ways in which economics and public policy affected that structure. The industry's structure determined how well it served the needs of electricity users. To an extent matched by few others, the electric utility industry has frequently been at the center of public policy controversy, and changing public policies have dramatically changed its structure. The time period covered begins with the roots of the industry that preceded Edison, continues through the industry's infancy and adolescence, and ends at a point where it had become an indispensable part of the lives of most Americans, the beginning of World War II. For many Americans living in rural areas, however, access to electric utilities still lay in the future.

Economics, Electricity, and Controversy

The economics of producing and distributing electricity are very distinctive; some are shared with a few other industries (such as railroads or other utilities), but no other industry shares them all. These characteristics combined with electricity's importance to create for the industry frequent, and sometimes sharp, political and policy controversy. Twice prior to World War II, these controversies led to national political movements that radically altered the industry and the environment within which it operated. Differences of opinion about the industry became ideological and hardened. Dispassionate objective analysis was nearly impossible. Making sense today of those controversies and their profound effects on the industry's development requires a basic understanding of the unique economics of electric utilities.

Public policies responded to concerns over specific negative aspects of the industry's behavior. They often were at least somewhat effective at addressing those concerns. However, the changes they created in the industry's structure inevitably had negative, unintended consequences that often led to new political controversies. The electric utility industry was never "fixed," although the intensity of the controversy sometimes temporarily abated. Few if any other American industries have had their institutional structures made and remade by public policy to such an extent, and the experience with electric utilities shows how difficult the design of economic institutions can be.

Decisions, Incentives, and Industry Structure

The many decisions made by individuals in an industry determine how that industry develops and responds to challenges. Most decisions seek outcomes for the decision maker that are likely to be rewarded with such things as higher pay, promotion, or greater prestige or esteem. Some of the incentives that guide decisions may encourage behavior that furthers the interests of the firm; others may not. Those decisions may or may not enhance the interests of the industry's customers and of society. Individual firms operate within an industry, and the institutional or organizational structure of the industry determines the incentives and constraints faced by firms and their decision makers. The structure of the electric utility industry is determined by the number of firms, the sizes of those firms in terms of both output and the geographic area served, the forms of firm ownership, and the relationships among firms in the industry and between them and firms in other industries. It defines the nature of competition between firms in the industry and with firms in other industries. In addition to its economic characteristics, the structure of the electric utility industry has been determined by the legal and regulatory environment of the industry, and that environment is largely the product of public policy.

A good institutional structure successfully aligns the incentives faced by firms and decision makers with society's needs so that when those decision makers act in their, or their firm's, interests their actions have a positive effect on their customers, on the economy, and on the overall public interest. An even perfect alignment does not assure that all decisions will be socially desirable; some may be unwise or based on bad information. A perfect alignment is not possible, but a poor alignment causes even the wisest and most informed decisions to produce socially undesirable outcomes.

No actual industrial structure can create ideal incentives. However, the theoretical structure of an industry in which many firms compete for the same customers, and into which new firms can easily enter, has served as a touchstone for industry structures. The economic or technological characteristics of an industry may hinder the ability of an industry to have this structure. For example, economies of scale may mean that only a few large firms can efficiently serve all of an industry's customers, and it may be very difficult for new firms to become established. Even then, the ideal of many firms competing remains a public policy standard. For example, antitrust actions that result in breaking up a large firm into several smaller ones are designed to bring an industry's institutional structure more in line with the competitive ideal. Controversy over such an action may hinge on the question of whether the smaller firms will be as efficient as the larger and whether the benefits of competition will outweigh this loss of efficiency. The desirability of competition is rarely controversial. However, the economic and technological characteristics of electric utilities prevent this ideal from being applicable to their industry.

Private Ownership, Government Ownership, and Competition

It has never seemed possible for electric utilities to operate in an industry structure in which the users of electricity could effectively choose among many competing firms for their supply.[2] Not long after the beginning of the twentieth century a consensus arose both that electric utilities had to operate as local monopolies and that the threat this posed to their customers justified an unusual level of continuous government involvement. With competition off the table, there was little agreement about what institutional structure for the industry would be best. For decades, the controversy hinged on the question of whether continuous government regulation of profit-seeking privately owned firms in the industry could effectively counter the undesirable incentives facing a protected monopoly. One side argued that such regulation would be effective and that private ownership encouraged positive behavior even in the absence of competition. The other side held that the political and economic power of privately owned electric monopolies would undermine the effectiveness of regulation to protect the public interest and that electric utilities needed to be owned and operated by government.

The lack of competition was not the only factor mitigating a direct role for government in the industry; the importance of hydroelectricity provided others. The largest and most important hydroelectric sites often were on public land. The construction and operation of hydroelectric sites often had implications for other activities for which government involvement was widely accepted, including river navigation and the provision of water for irrigation in arid areas. Hydroelectricity provided the seeds of what became an extensive role for government, particularly the federal government, both as a participant in and as a regulator of firms in the industry.

The struggle over the proper firm ownership constantly dogged the industry during the entire period covered by this book. The arguments made were often shrill and deceptive. Enormous resources were dedicated to influencing public attitudes and policy, especially by the privately owned segment of the industry. Private ownership has always dominated the electric utility industry, but the role played by government-owned utilities has always been significant. The electric utility industry is unique among large industries in the United States in containing this mix of firms owned privately and owned by various levels of government, and this is a legacy of that persistent historical controversy.

The Importance of Capital Costs

Capital equipment, the long-lived physical items required by an electric utility, includes generating equipment, transmission and distribution lines, land, buildings, and all of the other equipment needed to handle and sell electricity. Capital equipment is needed by all businesses, but for electric utilities, their costs have been unusually important, and much of the most important equipment has unusual and even unique economic characteristics. Both their costs and characteristics profoundly affected the industry's development.

For much of the electric utility industry's early history, the relative cost of capital equipment far exceeded that of any other contemporary industry. Until 1915, each year the industry spent more on capital equipment than it received in total revenue. To obtain the needed funds, the industry depended crucially on outside investors. This need had major effects on the industry's structure and on its relationship with the investment banking industry, and these effects are shown in multiple chapters, particularly chapters 2, 3, and 5. In the late 1920s, the increasing role of investment bankers in the industry contributed to the greatest political crisis and public policy the industry experienced during the period covered by this book.

The Peak-Load Problem and Capital Costs

The customers of electric utilities buy energy; electric energy is measured in kilowatt-hours (kWh). The cost of a utility's capital equipment, however, is not determined by the amount of kilowatt-hours produced but by the maximum amount of power (kilowatts or kW) that must be provided at any point in time, a time referred to as the system "peak."[3] 100 kW of power generated for a single hour produces 100 kWh of energy. One kW of power generated for 100 hours also produces 100 kWh, but the cost of the capital required for the former is much greater than for the latter. Electricity must be generated at the same instant it is used, and that use fluctuates over time. During periods in which the use of electrical energy is low, such as the middle of the night, no additional capital equipment is needed to generate more energy. During the system peak, however, any increase in energy use will require additional expensive capital. It is far more expensive for an electric utility to increase the amount of the energy produced then than when usage is low. Most produced goods do not face this situation because they can be stored. Even if the use of those goods varies, their rate of production can be held constant. Although electricity storage is possible, it has been, and remains, very expensive.

Joint Costs, Nonstorability, and Price Discrimination

Much of the capital equipment used by an electric utility, including such things as generating equipment and transmission lines, simultaneously serve all customers. The institutional structures of the industry have generally required that these joint costs be divided among all classes and subclasses of customers (e.g., residential, commercial, industrial) so that the prices charged a customer would purportedly cover the costs of serving just that customer or customer class. This, however, is impossible to do in an objectively correct manner and is not done by firms competing in a market. For example, a steer provides a number of different products: different cuts of meat, leather, gelatin, and more. The costs of raising a steer are joint costs; they simultaneously serve all of these products. A farmer, however, does not attempt to allocate those costs among the various products. The prices received for those products are determined by market demand and supply; those prices must provide the farmer with sufficient total revenue to pay the total costs of raising the steer.

The users of electricity can neither store it nor sell it to others. This situation enables price discrimination, where prices are set not according to costs but according to the demand characteristics of the customer. Changes in prices affect how much electricity a user will purchase, but users will differ in exactly how much their usage will change if the price changes. Under price discrimination, those whose usage is relatively less sensitive will be charged higher prices than those whose usage is more sensitive. This can enable a utility to earn higher profits than it would if all users were charged the same price. There is a positive side to price discrimination, however. There are situations, including ones that likely were present in the electric power industry, where price discrimination was in the public interest and might have enabled lower prices for everyone, a point discussed in chapter 3. Despite this, price discrimination contributed to the industry's political problems, since those charged higher prices are likely to view them as unfair. The lack of a correct way to distribute joint costs enabled methods of doing that to be used that supported price discrimination. Unfortunately, those methods were often maintained even after changing conditions could no longer justify the price discrimination they supported.

Transaction Specificity and Sunk Costs

The capital facilities required to generate and distribute electricity have little value if they cannot be used for the purpose originally intended; their costs are "sunk." For example, a utility's distribution system has no other use than to distribute electricity, and it cannot be moved to perform that function in another location. This contrasts with capital equipment that can be easily used for something else (such as land) or moved to another location. Although this may not allow recovery of all of the original cost, it would enable recovery of a significant portion. To remain in business, the revenue a utility receives must, over the life of the capital equipment, cover its costs, that is, repay those investors whose funds went to buy the equipment. Heavy dependence on transaction-specific capital equipment, however, will cause a business to continue to operate for a while even if its revenue is too low to cover those costs fully. Ceasing operation would cause all of its transaction-specific capital equipment to become worthless; none of their costs would be recovered. If revenue at least covers bare operating expenses, any amount remaining after paying those expenses could be used to recoup some of the capital equipment costs.

The concept of transaction-specific costs was developed to provide a partial explanation why some economic transactions occur between firms in a market while others occur within a single firm, perhaps between two divisions of the same firm. This can help explain the size of firms in an industry and the extent of their vertical integration. The need for investments in transaction-specific assets creates a risk for interfirm market transactions not present if the transaction occurs within a single firm. For example, suppose one firm purchases an expensive piece of capital equipment whose services will be jointly used by several firms. Before this is done, all of the firms must agree on the division of those services among the firms and the prices paid for those services. For this to work, all of the firms must adhere to the agreement. Once the capital equipment is in place, however, the bargaining strength of all the firms change. One firm might then take the position that it cannot pay as much as was originally agreed. The transaction-specific nature of the equipment means that it will still provide those services even if its costs can no longer be covered. Long-term contracts attempt to anticipate all contingencies, but this is never possible. The chance that one of the firms will claim an unforeseen contingency and break the agreement creates a risk that is avoided if both the investment in this type of capital and all of its use are made by the same firm.[4] The transaction-specific nature of much of the very expensive capital equipment required by an electric utility affect the industry's structure in multiple ways. Chapter 3 explains the important role this characteristic played in the development of state regulation. Chapter 4 discusses how this characteristic of capital equipment prevented the industry from taking full advantage of large fully integrated networks. It has not only proven difficult to develop an industry structure that overcame this problem; public policy changes to that structure often made the problem worse.

Large Networks

A large fully integrated network provides multiple benefits over the use of several unintegrated, or only slightly integrated, networks covering the same geographic area. During the period covered by this book, the size of the most efficient generator was continually increasing. A large network might be able to use, or make better use of, an efficient large generator than a small network. A large network requires less total generating capacity than multiple small networks. It enables greater flexibility in the siting of equipment. For example, there can be significant benefits to locating a steam-generation plant where there is access to fuel transport and cooling water. When hydroelectricity is involved, a large network provides additional benefits, particularly when it enables control of many hydroelectric facilities on a single river system.

Two factors enable one large network to require smaller total generating capacity than multiple small networks serving the same area. In an area with multiple unintegrated networks, the total generating capacity is determined by the sum of the peaks of each small network. Unless all of those peaks occur simultaneously, the peak of a larger network will be less than that sum. Different mixes of user types and variations in weather and geography can contribute to peaks occurring at different times in different areas. Because generating equipment is subject to unplanned breakdown, maintaining reliability requires that some generators be held in reserve to be used when such a breakdown occurs. The more generators on an integrated network, the lower the proportion of total generating capacity that must be held in reserve to meet any target level of reliability.[5]

Most hydroelectric facilities are located in areas where water flows vary, both seasonally and randomly, depending on the rates of rainfall or snowmelt. This means the amount of hydroelectricity that can be generated also varies. For a user of electricity, reliability of supply is very important, and electricity that can be reliably provided is much more valuable than electricity that is sometimes unavailable. Hydroelectric facilities often divided electricity into three categories. "Firm" electricity was always available and sold for the highest price, but the amount of firm electricity that could be produced was determined by the lowest water flow. "Interruptible" power was available only when generation exceeded the amount of firm power. It sold for much less. Finally, "dump" power was generated when water levels were unexpectedly high, and the reservoir's level was near its maximum. That water had to be let through the dam whether or not it generated electricity. Its value would be low since it could not be easily planned for, and any price would be better than not generating it.

A large fully integrated network can the increase the value of hydroelectricity by increasing the proportion that can be sold as firm power. One way of doing this is to have steam generators on the same network that can "firm up" the hydroelectricity by generating during times when water flows are low. The coordinated operation of multiple dams on a single river system can also increase the proportion of firm power. A tributary dam can hold water in its reservoir during periods of heavy rainfall and release it during periods when water flows are low. This will increase the flow of water on the main river and increase the proportion of firm power generated by dams on the river. If the tributary dam also has generators on the same network, it can add the electricity it generates to that of the main river dams, further increasing total generation during low water flows. A tributary dam not on the same network would operate differently. It would use its reservoir to even its generation, but it would generate at all times, not just during periods of low water flow. Generation on the main river would be more variable and the value of electricity from all dams would be less. This characteristic of hydroelectricity will be seen in chapter 7 where, among other things, it contributed to the role of the federal government's role as a generating utility.

Transmission

Technical and economic aspects of the transmission of electricity pose an additional problem to the development of a fully integrated network under multiple ownership. The transmission lines of a network are frequently interconnected, resulting in multiple paths between a generator and a user of electricity. In such a network, electricity flowing from generator to user concurrently uses all possible paths. With multiple generators and multiple users on such a network, a single transmission line will simultaneously carry electricity from many or all generators to many users. The owner of a transmission line has no control over the use of that line by any single generator or user and cannot even determine the amount of electricity from any single generator that used the line. This is very different from other types of networks, such as switched telephone networks or the Internet. Both sides of a telephone conversation can be determined, as can the origin and destination of a data packet travelling over the Internet. Any owner of a segment of the telephone network or the Internet can require compensation from those using the services provided by that specific segment. When only a single path is used, but multiple paths are available, competition is possible among the owners of the different paths. By contrast, the owner of an electric transmission line cannot be compensated by those benefitting from that line. Furthermore, decisions made by anyone on the network can have either a positive or a negative effect on the transmission capabilities of all other lines. These conditions make it difficult to provide multiple owners of existing or planned transmission lines with incentives encouraging overall network efficiency, even though the engineering problem of improving network efficiency can be solved. This problem disappears if the entire network is under single ownership. The problems of multiple ownership of a transmission network are discussed in chapter 4 and in the conclusion.

Common ownership of both transmission and generation solves another problem. If the need for electricity on a network increases, that additional need might be met most cheaply with additional generating equipment, additional transmission equipment, or a combination that depends on the specific situation. If transmission and generation are separately owned, creating a system of incentives that will result in the best possible mix is problematic.

An understanding of economic principles would not have enabled one in the 1890s to predict the evolution of the new electric utility industry. As with any aspect of human history, contingent events played a major role as did economic, technological, and political developments that originated outside the industry. A full story of how the industry developed must consider the effects of all of those events and developments. The special characteristics of the economics of electric utilities, however, helped determine how the industry responded to its changing circumstances, whether they occurred during its infancy, adolescence, or once the industry had become an indispensable part of American life.

Early Commercialization

Static electricity was known in ancient times as a property of amber, which, when rubbed with fur or wool, will take on a static electric charge sufficient to attract small bits of material. The word for amber in ancient Greek (and fifteenth-century English) was *elektrum*. Interest in this property lay dormant until 1600 when William Gilbert, royal physician to Queen Elizabeth and King James, published *De Magnete*, in which he coined the term "electric" to refer to this static electric property. Otto von Guericke, in the 1650s or early 1660s, invented the first electrostatic generator, which produced static electricity by friction on a revolving ball filled with powdered sulfur.[1] In the 1740s, the development of the Leyden jar, a type of capacitor, enabled the storage of relatively large amounts of static electricity for days, which could be discharged at will. It was also in this era that Benjamin Franklin contributed to the scientific understanding of electricity by, among other things, demonstrating its presence in lightning.

Static electricity generally has very high voltage but low current and power. In 1800, the Italian physicist Alessandro Volta invented the battery, and early improvements made available electric currents with higher current and power. Very soon after Volta's discovery, the English scientists William Nicholson and Anthony Carlisle discovered electrolysis, the use of electricity to separate compounds into their elemental components, by separating water into hydrogen and oxygen. By 1809, their compatriot Sir Humphry Davy first isolated a large number of elements by using batteries that could provide the high current needed. Commercially important electrochemical processes, including electroplating, and the smelting and refining of many materials soon followed. Two methods of producing artificial light from electricity, arc and incandescent lighting, each became a major impetus for the commercialization of electricity. Sir Humphry Davy demonstrated electric incandescent lighting in 1801 by using batteries to heat platinum strips hot enough to glow. There is some evidence that Davy demonstrated arc lighting the following year, but in 1808, he provided a well-documented spectacular demonstration of arc lighting to the Royal Institution powered by 2,000 battery cells, perhaps the brightest artificial lights heretofore seen.[2] In 1820, Hans Christian Ørsted discovered that a magnetic field surrounded a wire conducting electricity. The following year, Michael Faraday, a onetime laboratory assistant to Davy, made what was probably the first electric motor.

Devices that could provide ever-larger quantities of electricity stimulated the scientific study of electricity. Batteries provided sufficient electricity to enable the commercialization of some electrochemical processes, as noted above, and to support the transmission of information by telegraph and telephone. For other applications, including light and power, the use of electricity remained a curiosity until the development of the generator substantially reduced the cost of producing large enough quantities of electricity to make possible an electric utility industry. The generator also enabled the creation of alternating current. Although batteries only produce direct current, the most common design of a generator more naturally produces alternating current. A simple device, the commutator, can convert this to something approximating direct current. Some uses for electricity, however, such as electroplating, required direct current. The development of both direct and alternating current generators occurred very early. The special properties of alternating current were to have a profound effect on the development of the electric utility industry, but it took some time before these properties and their importance were understood.

The first generators used permanent magnets. The discovery of self-excitation, where permanent magnets were replaced by electromagnets using current produced by the same generator, led to substantial improvements in generator design. Generators with permanent magnets were called magnetogenerators, and those using self-excitation were called dynamogenerators, usually shortened to "dynamos." Since electromagnets require direct current, a true dynamo can only generate direct current. Electromagnets were soon used in alternators, however, and these also were called dynamos, although the electromagnets were usually powered by a separate direct-current generator. In 1871, Zénobe-Théophile Gramme, who was born in Belgium but did much of his work in France, produced the first commercially successful dynamo. Although until the late 1870s its use was primarily by the electrochemical industries, it made the more widespread use of electric lighting inevitable, and by 1879 Gramme had sold over one thousand of them. Gramme's dynamo could also function as a motor, as he demonstrated at the Vienna exhibition of 1873. In the Philadelphia centennial exhibition of 1876, Gramme dynamos powered arc lights and electroplating demonstrations and also ran as motors.[3]

Light

Artificial lighting was the first application to drive the development of the electric utility industry. The modern utility industry sells electrical energy for all purposes, a characteristic that distinguishes it from, among others, the telephone and telegraph industries that sell a service produced with electricity. The industry that became the electric power industry began also by selling a service produced with electricity: artificial light. Artificial light required more power than telegraphs or telephones, and the superiority of electric lighting over other forms of artificial light led to the need for an infrastructure capable of distributing large amounts of electricity.

It is difficult to appreciate fully life without satisfactory artificial lighting. Modern calendars sometimes provide the phases of the moon, now a quaint decoration. As late as the beginning of the nineteenth century, however, knowledge of the phases of the moon was of significant practical importance, because most people simply did not venture out at night unless there was adequate moonlight.[4] Prior to electric light, artificial light required a flame from such devices as a fire, a candle, or an oil or gas lamp. These generally suffered from major disadvantages compared to the electric light: (1) the quality and amount of light was low, (2) they created heat and soot, (3) they consumed oxygen, (4) they smelled, and (5) they generally required constant maintenance. We may today use candles or oil lamps when the electricity fails or to achieve a romantic atmosphere, but the modern candle and oil lamp themselves are enormous improvements over the candles that were available until the end of the eighteenth century.[5] Prior to electric lighting, the desire for improvement in artificial lighting was a major factor driving the whaling industry and the infant petroleum industry, both of which provided products that substantially improved the quality of candles and lamps.

Gas lighting, which became available in metropolitan areas in the first couple of decades of the nineteenth century, was a considerably improved form of artificial lighting. A central plant would heat an organic material (usually coal) in the absence of air. This would produce a gas then piped to customers in the area. Those customers burned the gas in a fixed lamp or other appliance designed for this purpose. Like other forms of lighting, the

This Room Is Equipped With

Edison Electric Light.

Do not attempt to light with match. Simply turn key on wall by the door.

The use of Electricity for lighting is in no way harmful to health, nor does it affect the soundness of sleep.

FIGURE 1.1. Edison sign. Digitally enhanced to improve clarity.

light came from an open flame.[6] Designs for gaslights enabled them to be brighter than most other forms of artificial lights by using multiple flames and larger flames than candles or oil lamps. Perhaps the major advantage of gas lighting was that it required much less maintenance than other forms of open flames. There was no need to refill a fuel reservoir, and there was no wick to constantly adjust and trim. For these reasons, gaslights enabled the more widespread use of artificial outdoor lighting, such as streetlights. Nevertheless, gaslights retained all of the other disadvantages of open flames, and electric utilities touted the advantages of electric light (figure 1.1).

Humphry Davy demonstrated the principles by which electricity could produce light: incandescent and arc light. In fact, both techniques used incandescence—heating a material until it glows—to produce light.[7] In an incandescent light, electricity flows through a conductor whose resistance causes the conductor to heat to incandescence. In an arc light, a spark crosses a gap between two electrodes. This spark heats the electrodes (particularly the anode, or positive electrode). Although the spark itself produces some light, most comes from the incandescing anode. The light from an arc light is very intense—too intense for most indoor applications but well suited to outdoor illumination and the illumination of very large interior spaces, as in public buildings.

The primary problem that both arc and incandescent lighting had to overcome to achieve commercialization is that the heat required for incandescence tends to destroy the incandescing material. This destruction eventually creates a gap in the incandescent conductor, which stops the flow of current and extinguishes the light. In an arc light, the destruction of the electrodes occurs only in the direct vicinity of the arc, but this increases the gap width until it becomes too great for an arc to span, also stopping the current and extinguishing the light. Attempted solutions to these problems inevitably brought forward other problems. The problem of incandescent lighting was solved once it became possible to seal the conductor in a glass bulb from which oxygen had been eliminated, reducing the rate of deterioration of the incandescing conductor. The most common solution to the problem faced by arc lighting was to devise a "regulator" that would automatically adjust the interelectrode gap. The earliest regulators limited a generator to powering only a single arc light, and full commercialization waited until this problem was overcome.[8]

Arc Lighting

The Russian telegraph engineer Paul Jablochkoff, who worked in Paris, designed the first commercially successful arc lighting system in 1876. Jablochkoff's "candle" eliminated the need for a regulator by placing the electrodes in parallel with a solid material used as a spacer. The lamps were cheap but short-lived and could not be relit once they were turned off. Jablochkoff was able to install several lamps in series in a single circuit, and he was able to make lamps of varying brightness (although all were too bright for residential interior use). The connection of arc lights in series became the standard industry practice. A drawback to this arrangement is that an interruption in the flow of current through any device would shut down the entire circuit, plunging the whole area served by the circuit into darkness, although various solutions to this problem were eventually developed. Series connection permitted a single switch to control an entire circuit of lights, a desirable feature for streetlights and the lighting of some large interior spaces.

The problem of maintaining an arc lighting circuit when one of the lights burned out brought Jablochkoff to the threshold of discovering the principle of the transformer, a device that was to totally remake the infrastructure of electric utilities. To ensure that both electrodes were consumed evenly, Jablochkoff employed alternating current in his system. Alternating current (but not direct current) produces a varying magnetic field that can induce current in adjacent conductors, this principle being the basis of the transformer. Jablochkoff experimented with the use of induction coils, actually transformers, to connect each individual lamp to the circuit. Connecting an arc light to its circuit via an induction coil enabled an arc light's failure not to stop the flow of current in the circuit. However, he found methods of dealing with the problem of interruptions in the series circuit preferable. Jablochkoff's system was installed in numerous locations in Paris both by his Société Générale d'Electricité and by others and, in 1878, in various locations in London.[9] Ultimately, however, systems that used lamps with superior regulating mechanisms superseded Jablochkoff's.

The more lights, or other electricity-using devices, on a circuit, the greater the power that circuit must supply. Electrical power is the product of voltage (pressure) and current (amperage). Modern electric utility circuits maintain constant voltage and vary amperage as needed to adjust total power to the needs of the devices on the circuit. The original arc lighting circuits, by contrast, maintained constant amperage and varied voltage when the number of arc lights on the circuit changed.[10] A characteristic of arc lights is that their resistance is inversely proportional to the amperage of the current flowing through them; higher amperage results in lower resistance. In such a circuit, a constant voltage generator would tend to increase amperage whenever an arc light's resistance dropped. The increased amperage would further reduce the light's resistance, leading to even greater amperage. High enough amperage could cause the circuit to overheat, potentially resulting in a fire. Constant amperage generators avoided this problem but led to the need to design a very different circuit for incandescent lighting, one that required protection, such as fuses, to protect against excessive amperage.

In 1879, the first arc lighting central station, the California Electric Light Company of San Francisco, began operations. It is an indication of how rapidly technical events were progressing in the industry that in the same year a workable incandescent bulb was demonstrated by Joseph Swan in England (in February) and Thomas Edison in the United States (in October). Although a number of other inventors had also worked on incandescent lighting, and were soon to develop practical designs, Swan and Edison are the ones most often identified with the invention of the incandescent bulb.[11] The intense brightness of arc lights made them unsuited for the majority of situations where artificial light was needed, such as inside residences. Incandescent lighting had long been seen as a possible solution to this problem, and Swan had begun working on incandescent lighting as early as 1848—before generators or high vacuum pumps were available and while Edison was still a teenager. In 1881, the incandescent lamps of four different inventors or partnerships were exhibited at the first international electrical exhibition in Paris.[12]

In America, at least, Edison is widely known as the "inventor" of the incandescent light. This is inaccurate and fails to recognize the actual fundamental contribution he made to the development of electric utilities—one that rendered obsolete the entire technological foundation of the nascent arc lighting industry. A person of enormous personal inventive genius, Edison was also a skillful organizer and promoter. By the time he began his work on electricity, Edison had a staff of forty people, including the mathematically trained Francis Upton. Edison was not a solitary inventor but more like a symphony conductor orchestrating the effort of his entire staff, and he had a much broader conception of the problem of electric lighting than did Swan. The central problem that concerned Swan was that of developing a method to make a durable incandescing conductor. Although mindful of the problem of durability, Edison developed an entire technical basis for a wholly new electric lighting industry. He then proceeded with his staff to invent all of the components for this new industry, of which the light bulb, although key, was only one element.

Edison took gas lighting, not arc lights, as his model for a new incandescent lighting industry.[13] Like gaslights, Edison wanted incandescent lights to be independently controlled. This meant that the lights would have to be connected to the generator in parallel, not in series, and this change profoundly affected the electrical characteristics of the system.

The relationships between the power used by an electrical device, and the voltage and amperage of the circuit are shown by the following two equations, both derived from Ohm's law:

$$P = VI = I^2R$$

and

$$I = \frac{V}{R}$$

where P is power, V is voltage, I is amperage, and R is resistance.

Every conductor, including those used in the transmission and distribution network, has some resistance. That resistance will cause some of the electricity that passes through it to be dissipated as heat. Unless this occurs in an appliance where the production of heat is desirable, it wastes power. The higher the amperage, the greater the power converted to heat (and wasted) by the distribution system and the greater the heat produced. If allowed to become high enough, this heat can destroy the conductors and create a fire hazard. The first equation shows that minimizing the power consumed by a conductor in the distribution circuit requires reducing the total amperage in the circuit or the resistance of the conductor, or both. Reducing the resistance of a copper wire requires increasing its diameter and thus its cost.

If connected in series, as was the practice with arc lights before Edison, the same amperage in the circuit is available to each device, regardless of the number of devices, and the size of the required conductors did not change when the number of lights was increased. The number of arc lights in the circuit, however, determined the total resistance of the circuit. The relationship given by the second equation shows that in order to maintain the same amperage, an increase in resistance required the voltage of the circuit to increase. Adding arc lights to a circuit increased the circuit's total resistance, and the constant-amperage generators responded by increasing voltage.

Parallel circuits, required for individual lamp control, have very different electrical characteristics than series circuits. In a series circuit, increasing the number of lights in the circuit does not affect the required amperage. By contrast, in a parallel circuit, the circuit's required amperage is the sum of that required by each light. As the number of lights increases, so must the amperage and the cost of conductors. The amount of light produced by a light (arc or incandescent) depends on the power it uses. The first equation shows that a bulb using low amperage must have high resistance for high power. Edison recognized that the design of an incandescent light required its electrical resistance to be as high as possible. This required that the incandescing conductor (filament) in the bulb be as thin as possible, a feature that adversely affected the bulb's durability. Swan's initial incandescent bulb was designed to replace an arc light in a series circuit. Like an arc light, its thick incandescing conductor had low resistance and required high amperage. Edison overcame the greater challenge of designing a bulb for the parallel circuits he envisioned. Parallel circuits quickly became the standard for incandescent lighting, and all bulbs, including Swan's later bulbs, had high resistance.

In addition to imposing a constraint on bulb design, Edison's parallel circuits required many new components that were developed by his organization. These included new wiring techniques, new constant-voltage generators, meters capable of measuring an individual customer's use of electricity, and fuses that limited the maximum amperage of a circuit. The list of Edison's inventions for his new system in 1881 occupied five full pages in the patent office's *Official Gazette*.[14] Edison's competitors invariably adopted his concept of electricity distribution and developed solutions to the same problems Edison anticipated. While others can compete with the claim that Edison invented the incandescent bulb, he was the first to conceive of the wholly new system into which the bulb would operate, he was the first to develop a bulb capable of operating in that system, and he was the first to implement such a complete system. In 1882, an experimental Edison station was established in London, and, a few months later, the first permanent electric utility designed to serve incandescent lighting was built by Edison on Pearl Street in New York City.

A number of factors enabled competitors to enter the industry Edison had created. A newspaper published technical details of his system in 1879, several years before he was ready to put them into practice, and this allowed others to begin designing their own systems. Until 1885, Edison's partners were not vigorously suing those who might have infringed his patents. Edison faced other problems. A new electric utility required an extremely large investment before it could begin operation. Edison's backers were not just reluctant to invest in electric utilities; they hesitated to invest even in the manufacturing of the equipment those utilities would have to purchase. Edison was able partially to finance this necessary manufacturing by selling his control over the basic patents and combining the funds he received with additional outside backing. This resulted in a number of legally separate companies all bearing the Edison name and all involved in manufacturing or marketing some aspect of the new technology. Friction between Edison and his original backers increased since they, rather than he, now controlled the basic technology. Edison's situation eventually improved, culminating with the merger in 1889 of the different Edison enterprises into the Edison General Electric Company by the financier Henry Villard. However, the inability of Edison to overcome the challenge from alternating current made the new company short-lived. In 1892, over his objections, Edison's company was essentially taken over by a manufacturer of alternating current equipment, the Thomson-Houston Company, creating the modern General Electric. Edison withdrew from active involvement in the industry he had done so much to create and with which he would always be associated. Instead, he devoted his considerable inventive and organizational genius to new and undeveloped areas.[15]

The First Electric Utilities

As more companies became involved in the manufacture of generating and other electrical equipment, they needed a growing pool of customers. There were two categories of customers of the new electrical equipment manufacturers: electric utilities that sold electricity to many users and those whose intent was to produce electricity solely for their own use in "isolated plants." The latter group initially dominated the demand for electrical equipment and remained important for decades. This group included factories that generated the electricity needed for operations, hotels and retailers that generated electricity primarily for lighting their own buildings, and municipalities that generated the power required for lighting streets and other public places. Transit companies that used electricity to power streetcars or light railways would often produce the electricity they needed. Even wealthy individuals willing to pay to be on the technological frontier would install generators in their homes. Electric utilities or "central stations" were both privately owned and owned by governmental entities—initially municipalities.[1] The equipment required to efficiently generate electricity was always much larger than the needs of many who wanted to use it. This created an economic niche for utilities, enterprises that could produce electricity on a large scale and provide it to those with relatively small demand. Over time, the most efficient scale for electricity production increased, even relative to the growing average demand for electricity by users, and this increased the need and opportunity for electric utilities.

Paying for Capital Equipment

Electric utilities were to become the major customers for the manufacturers of electrical equipment, but first they had to be created. The economic problem facing the incipient utility was the necessity of somehow raising the enormous amounts of money required to purchase the equipment and construct the infrastructure before it received any revenue. The problems of raising capital, and the consequent need for large numbers of outside investors, greatly influenced the growth and structure of the electric utility industry.[2] The early manufacturers of electrical equipment also faced the problem of raising funds. The first residence Thomas Edison provided with incandescent lighting was that of the financier J. Pierpont Morgan in 1881.[3] The following year he located his first permanent central station on Pearl Street in lower Manhattan, where he could provide electricity to the nation's financial center. These were strategic decisions that enabled him to give early access to the new technology to the very people on whom he would need to depend. The new utilities faced an even greater struggle. Investors in the electrical equipment manufacturing industry had reason to feel concern about a class of customers who themselves were starved for financing. By contrast, customers in a position to pay cash up front were the ones who would bring a quicker return. It was the second group of customers, those producing electricity for themselves, not the utilities, that equipment manufacturers saw as desirable.

Nevertheless, the potential for privately owned electric utilities to become major customers was one that American electrical equipment manufacturers recognized early. The pioneer arc-lighting entrepreneur in America was Charles Francis Brush, who developed improvements both in the design of arc lights and in the dynamos that powered them. Although Brush initially sold complete systems, including dynamos and lights, to individual companies (such as stores), he saw a large potential market in street lighting, whose development led to the first central-station electric utilities. A number of American cities were then using gas for street lighting, and private utilities that supplied gas for both private and public (street lighting) use were the typical providers, not municipally owned utilities. Brush, like every electric utility promoter after him, had to deal with the financing needs of the new local utilities. The magnitude of the required financing made it very difficult for an equipment manufacturer to create and operate local utilities itself, and Brush's own company concentrated on the manufacture of equipment, not the operation of local utilities. A Brush representative would go to a city to try to organize a utility financed and owned by local interests. This new company would enter into an agreement with the Brush company giving it exclusive territorial rights to purchase Brush equipment for utility use and entitling it to a commission on all Brush systems sold in its territory as isolated plants. The local utility was also required to give Brush between 32 and 48 percent of the stock of the local utility. In the summer of 1879, operations began at the California Electric Light Company of San Francisco, the world's first central-station electric utility, and other Brush utilities were set up in a number of American cities including New York, Boston, and Philadelphia.[4]

Brush's approach was one followed by other electrical equipment manufacturers: a local utility received an exclusive territorial franchise that extended beyond the utility business to include the sale of electrical equipment for nonutility use. In return the manufacturer would receive financial instruments—stocks or bonds—whose value would depend on the ultimate success of the utility. The acceptance of financial instruments caused the manufacturer to participate in the risk otherwise borne by the utility's investors. This participation was much greater, of course, if the manufacturer accepted stocks rather than bonds, in part because secondary financial markets for common stock were not well developed in the United States at the turn of the twentieth century.

One of the factors that encouraged the involvement of manufacturers in the early creation of electric utilities was the relative lack of standardization in the new utility industry. Those setting up a central station had to choose among several available systems for both arc lighting and, eventually, incandescent lighting. Once the managers of a new electric utility chose a particular manufacturer's system, virtually all of the components would have to come from that same manufacturer to ensure that everything worked together properly. The early electric utilities became a captive customer of the manufacturer. Both the manufacturers and the investors in electric utilities knew this, of course, and that affected the nature of the license contracts between them. Although Edison laid the basis for a "universal" system, the Edison companies tried to achieve this kind of "lock-in" with the first incandescent electric utilities. With the same business model in mind, Thomson-Houston delayed marketing any alternating-current equipment until Elihu Thomson had developed a complete system encompassing all of the components for the generation, distribution, and consumption of alternating current.[5]

The Edison interests followed Brush's policy in cities with an 1880 census population of 10,000 or more, the only cities in which it engaged in any risk sharing; central stations in smaller cities had to pay cash for their equipment. In the larger cities, the Edison franchise required the utility to give to the Edison company between 20 and 30 percent of its common stock, to pay a fee based on its capitalization, and to agree to use only Edison equipment. In return, the franchisee received an exclusive territorial right to operate a central station with Edison equipment and to sell all Edison lighting equipment.[6] The licensees were charged prices for equipment substantially below those paid by others. For example, prior to 1890, the standard price for an Edison light bulb was $1.00, but licensees paid $0.40.[7] Westinghouse, an early competitor with Edison, did not use a formal licensing system but used financial inducements to encourage central stations to agree to use Westinghouse equipment exclusively and imposed financial penalties on utilities that reneged.[8]

The manufacturers could have accepted even more risk (for greater potential reward) had they actually created and owned the utilities themselves. Such an approach would have imposed an enormous financial burden, however, and would have required the manufacturers to be more investment bankers than equipment manufacturers. Occasionally, a manufacturer would use a variant of this approach. A notable example was that of Edward H. Goff, who started as a sales agent for American Electric Company, an early manufacturer of arc lighting equipment designed by Elihu Thomson and Edwin J. Houston. Goff would own and operate local arc lighting utilities for three or four years before selling them to local entrepreneurs. Although he was quite successful initially, this method put an enormous strain on his company's finances that ultimately contributed to its downfall.[9] Westinghouse later used this method to start some electric utilities, and Edison favored this approach for smaller towns, but he was initially unable to convince his financial backers to agree to its use, although he eventually did set up utilities in twenty smaller towns.[10]

Government-Owned Utilities

Although the direct involvement of government-owned firms in the US economy has been rare, utilities in general, and electric utilities in particular, have been important exceptions. From the beginning, municipal governments got in the business of owning and operating electric utilities.

TABLE 2.1 **Characteristics of municipal and private stations, 1902**

Population of town or city	Number of stations		Percent municipal	Total generator capacity in kilowatts	Number of stations		Percent municipal
	Private	Municipal			Private	Municipal	
Under 5,000	2,043	671	24.7%	Under 149	1650	612	27.1%
5,000 but under 25,000	554	121	17.9%	149 but under 373	700	164	19.0%
25,000 but under 100,000	115	13	10.2%	373 but under 746	201	25	11.1%
100,000 but under 500,000	67	6	8.2%	746 but under 1,492	126	11	8.0%
500,000 and over	26	4	13.3%	1,492 but under 3,730	81	2	2.4%
Total all sizes	3,620	815	13.3%	3,730 and over	47	1	2.1%

Source: US Bureau of the Census 1905, 10
Note: The census report measured generating capacity in horsepower, which has been converted to kilowatts in the table.

There were two reasons why a municipality would have established an electric utility. One was ideology. By the end of the nineteenth century, the Progressive movement had become concerned with the potentially corrupting effect private business interests could have on government operations. Electric utilities were among the business interests reshaping the economic environment of the period. The usual absence of a competitor to whom the customers of an electric utility could turn help create a sense of vulnerability that fueled strong negative feelings toward privately owned electric utilities.[11] Private electric utilities required a franchise from the city government in order to use the public streets for their distribution system. As discussed more fully in chapter 3, this put them at the center of corruption scandals in many cities. One of the methods favored by some Progressives to deal with this problem was to have utilities owned and operated by government, although many also thought that government regulation of privately owned utilities could effectively handle the problem. Some thought the mere threat of government takeover was an essential component of effective regulation.

The second reason many municipal governments set up electric utilities was both more prosaic and probably more common: it was the quickest way their citizens could get access to centrally generated electricity given the special difficulties of setting up a privately owned utility in smaller towns. As shown in table 2.1, some municipally owned utilities were in large cities and had large generating capacity. Compared to privately owned utilities,

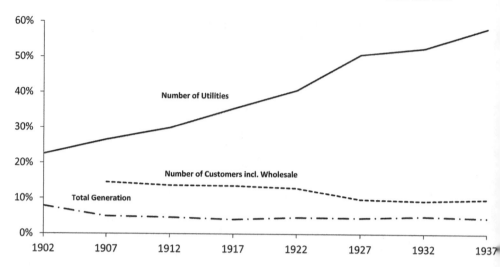

FIGURE 2.1. Municipally owned utilities' share of number of utilities, number of customers, and generation

Source: US Bureau of the Census, *Census of Electric Light and Power*, various years

however, they tended to be smaller and to be located in smaller towns. Their share of the total number of utilities grew but not their share of either generation or number of customers (figure 2.1).

The 1902 census collected from each station the year in which it began operation. These data are not ideal for determining the role of municipal stations in the early years of the industry because they fail to count both those stations that went out of business prior to 1902 and those stations that disappeared because of merger.[12] Nevertheless, these data can give some insight into the role that municipally owned utilities played in the initial formation of the industry, and figure 2.2 shows the proportion of electric utilities beginning operation each year prior to 1902 that was owned by a municipality. There was a clear tendency for that proportion to increase. In addition, 170 utilities that started under private ownership switched to government ownership while only thirteen utilities initially owned by the government switched to private ownership. The census report stated that these data illustrated in "a somewhat striking manner" the "agitation for municipal ownership of public service enterprises."[13] The long-running political controversy over the ownership of public utilities had already begun.

Perhaps the switch from private to government ownership in part reflected the difficulty of an electric utility making a profit in smaller communities, which led either to bankruptcy or to at least a desire on the part of owners to get out of that business. It is also possible that an additional impetus for municipal ownership was that prices for residential users, who dominated the electorate, would have been lower with municipal than with private ownership. However, evidence supporting this is lacking. As the nation gained more experience with electricity, the demand for it increased and the technology improved; areas with lower population density increasingly became attractive targets for the establishment of privately owned electric utilities. Between 1902 and 1907, the likelihood that a private utility would switch to municipal ownership was little if any greater than that of a municipal utility shifting to private ownership.[14]

Municipalities sometimes faced legal barriers in starting or acquiring an electric utility. Some states initially prohibited municipalities from operating electric utilities; others prohibited competition with any private company, required passage of a referendum before a municipal government could operate a utility, or severely limited the ability of a municipality to incur the debt required to acquire or start an electric utility. These restrictions, however, varied widely and were not present everywhere.[15]

Municipalities were not exempt from the basic economic problem facing anyone wishing to start an electric utility: the enormous up-front

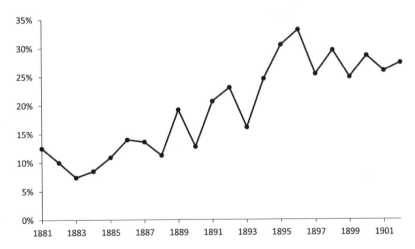

FIGURE 2.2. Municipal utilities as percentage of total, 1881–1901

Source: US Bureau of the Census 1905, 7

capital required. If a municipality was relatively free of legal and political constraints, it possessed an important advantage over newly formed private enterprises: better access to capital markets. A municipality had many assets, not the least of which was the power of taxation, which it could use to back any bonds it issued. A municipal utility shared an advantage with isolated plants over a private utility. Investors were more willing to loan money for electrification to a borrower that could back its debt with other assets. To be able to borrow money, a privately owned utility had to rely partly on equity financing, which was riskier, more expensive, and often harder to obtain than financing from the sale of bonds. By contrast, a municipality could finance a utility entirely from the sale of bonds.

The potential financing advantage a municipality had over a private company suggests that the costs to running an electric utility might have been lower for a municipality than for a private company in otherwise identical circumstances. In addition to these possible cost advantages municipal ownership had over private ownership, the earliest utilities owned by municipalities were much more efficient than were those privately owned.[16] There are many other reasons, however, why the prices charged by a municipal utility might differ from those charged by a private utility. Municipally owned utilities usually did not have the same tax liabilities as those privately owned. A municipal utility might decide for political reasons to subsidize the sale of electricity and charge a low price. An expectation of lower prices for residential users might have encouraged municipalities to start or acquire an electric utility, although evidence for this is lacking.[17] Conversely, a city might have decided to charge a high price and use the surplus from electricity sales to cover other city expenses or reduce tax rates.

Alternating Current and Large Networks

Although much of Edison's design for the industry infrastructure has endured, there was one very significant change—the switch from his direct current (DC) system to alternating current (AC). Alternating current made it easy to convert current with low voltage and high amperage to current with (almost) the same power but with high voltage and low amperage for transmission and then to convert that current back to lower voltage and higher amperage for use. The amount of power in an electrical current is a product of its voltage and its amperage; for current to carry a great deal of power, either the amperage or the voltage or both must be high. High voltage is undesirable because the difficulty in insulating it makes it dangerous and complicates the design of electricity-using equipment. High amperage is undesirable because current with high amperage produces heat in conductors, leading to current losses and posing a fire risk unless thicker and more expensive conductors are used.[18] Conductors that carry current a long distance must be thicker than those that carry it only a short distance. For long-distance transmission of electricity to be feasible, the amperage needs to be as low as possible, and this requires the voltage to be as high as possible.[19]

A moving or fluctuating magnetic field will induce an electric current in a nearby conductor. This same principle is used in all generators, where external motive power is used to cause a magnetic field to rotate relative to a coil of wire in which an electric current is induced. When electricity flows through a conductor it produces a magnetic field most effectively if the conductor is coiled, as in an electromagnet. Because of the fluctuating nature of alternating current, when AC current passes through a (primary) coil, it produces a fluctuating magnetic field. This fluctuating field can induce a current in a separate adjacent (secondary) coil without any movement in either coil or other change in the current. If the secondary coil has more turns than the primary, the induced current's amperage will be lower but its voltage will be higher. If the secondary coil has fewer turns than the primary, amperage will be higher but voltage lower. An alternating-current transformer is thus a relatively simple device. As technology advanced, the maximum voltage at which electricity could be transmitted continually increased, and this made feasible larger and larger networks that were unavailable to a DC system. As discussed in the introduction, these larger networks both reduced the total generating capacity required to serve an area and enabled the more efficient use of the generators that were used. This gave AC an advantage over DC that was insurmountable, although, as discussed below, the complete elimination of the use of DC by electric utilities took considerable time.

Edison's creative insight into the technological requirements of an electric utility industry is unquestionable, despite his failure at the outset to have realized that alternating current could offer benefits that made its use irresistible. It is, nevertheless, tempting to ask whether he should have or could have seen the advantages of alternating current before settling on direct current. Edison was surely aware of alternating current. Arc lighting

systems that used both forms of currents were in commercial service before Edison developed his system. As previously noted, Jablochkoff's arc
lighting system employed alternating current, and he even experimented
with the use of a transformer-like induction coil to handle the problem of
maintaining the flow of electricity in a series-connected circuit when one
of the arc lights failed.[20] Despite this awareness of the basic principle of the
AC transformer, it appears that prior to Edison's development of his system, there was no appreciation of the importance of a transformer's ability
to reduce, as well as increase, voltage.

Edison consciously modeled his system on the gas lighting system, but
an important difference between gas and electricity made the benefits of
large networks unique to electricity. Gas can be stored much more cheaply
than electricity; there is not the need for gas to be produced at the instant
it is consumed. Natural gas does require the use of long-distance pipelines,
but the capacity of these pipelines is dependent more on the average use
of gas over time than on the peak use at an instant in time. In Edison's
time, local plants manufactured the gas used, typically from coal, which
did not require long distance transport. His design of the high-resistance
incandescent bulb shows that he understood the importance of keeping
the level of amperage in a circuit low in order to reduce the required size of
conductors. In the face of the challenge from AC, he worked on methods
to increase the voltage at which DC could be transmitted and distributed.
He invented a three-wire system, still used today, that enabled electricity
to be distributed at twice the voltage at which it was used, but he was unable to devise a method for DC of easily changing a current's voltage and
amperage while maintaining its power.[21]

The utility of the transformer as a device both to increase and decrease
voltage was first recognized by the Frenchman Lucien Gaulard and his
English business partner, John Gibbs. Gaulard and Gibbs designed a
transformer with an adjustable secondary, which enabled varying the output voltage to according to the needs of the load. They first demonstrated
this transformer in 1883 and were apparently partly reacting to the British
Electric Lighting Act of 1882 that forbade suppliers of electricity from
prescribing the type of bulb that users could employ. There was then no
standardized voltage for electricity supply; different utilities supplied current at different voltages. A light bulb was designed for a single voltage;
the different voltage meant there were many different light bulbs available. A transformer capable of converting any voltage current to another
voltage enabled the use of any bulb on any system. Gaulard and Gibbs also

realized that another application for their transformer was a system that increased voltage for transmission and decreased it for use.[22] Their transformer design suffered from several problems, however. It was designed for use with the type of constant-current AC generator used for arc lights, and the primaries of the transformers were wired in series, as arc lights would have been.[23] This resulted in serious voltage regulation problems, and efforts to solve these problems added to the complexity of the system. The design changes required to deal with these issues were not difficult and required relatively little new technology: the use of constant voltage generators, the elimination of adjustability in the transformer secondaries, and the connection of the transformer primaries in parallel rather than in series. It was others, however, who made these improvements, including the American William Stanley, who was hired by George Westinghouse to help develop an incandescent lighting system.[24] During the period from 1885 to 1887, alternating current incandescent lighting systems moved from laboratory demonstration to full commercial operation in both the United States and Europe. Westinghouse used alternating current to illuminate the 1893 Chicago Columbian Exposition. The 1895 implementation of an alternating current system to harness the power of Niagara Falls (including transmission to Buffalo, roughly twenty miles) demonstrated the commercial viability of the system. Coal-poor European countries vigorously pursued long-distance transmission to connect distant hydro facilities with electricity users, and American manufacturers followed their progress closely.

The majority of users had little reason to prefer one form of current to another. Both could do an equally good job powering the same light bulb. Alternating current with too low a frequency caused a bulb to have an annoying flicker, but sufficiently high frequencies were easy to attain. For the utilities, however, the use of alternating current involved a completely different infrastructure: different generators, different regulating equipment, and, most important, a different-sized service area. These differences led to major changes in the way electric utilities operated, and the alternating current system immediately posed a serious competitive threat to the manufacturers of direct current equipment, especially Edison. Initially, direct current continued to possess a number of advantages. It took time before the vendors of alternating-current equipment could provide a total system as well developed as Edison's direct-current system. The first generators designed for alternating current incandescent lighting were not as efficient as were their direct current counterparts. Alternating current initially lacked meters and motors it could power. The first alternating motors that became available were inferior in applications that required high torque at varying speeds. Many applications, such as elevator motors, needed this capability. Direct-current systems had the advantage of being able to use storage batteries. Those batteries reduced the total generating capacity a utility needed and provided a backup if a generator failed.[25] Furthermore, many utilities had already made tremendous investment in direct-current infrastructure, especially in heavily urbanized areas. As a result, alternating current usually came first to areas without electric utility service—areas whose population densities were too low to be served economically by the transmission-constrained direct current system.

The "battle of the systems" (or "battle of the currents") between alternating and direct current that raged roughly between 1887 and 1892, had strange, even bizarre, aspects. Edison interests mounted a public relations campaign against alternating current based on the proposition that it was more dangerous than direct current. In several states, including New York and Virginia, they tried unsuccessfully to have the state legislature ban the use of high voltage alternating current.[26] Edison's West Orange, New Jersey, laboratory became the scene of grisly public demonstrations that purported to show the dangers of alternating current by using it to kill animals.[27] During this time, New York State decided to seek a more humane form of legal execution than hanging. Edison interests lobbied hard for electrocution using alternating current and, despite opposition by Westinghouse, managed to get the state to use a Westinghouse alternator for the first legal electrocution. This was followed by a publicity campaign in which alternating current was termed the "executioner's current," and an effort was made to term electrocution "Westinghousing." There is little evidence that these efforts to block the adoption of alternating current had much effect.[28]

One of the developments that accelerated the adoption of alternating current for generation was the invention of a variety of "gateway" technologies that permitted the conversion of alternating current to direct current for use. One such technology was the "motor-generator," which involved coupling a motor designed to run on one form of electricity with a generator designed to produce another. These were widely used to convert alternating current to direct current, but they could do many other types of conversion as well. A modified version of the motor-generator was the "rotary converter" with two windings on a single armature.[29] The rotary converter was initially more efficient but less versatile than the motor-generator.[30] These new technologies enabled utilities to use alternating current generation and transmission to supply direct current, but they did not enable utilities to have one generation and transmission system capable of supplying users with both direct and alternating current. Instead, a separate alternating current system provided each class of customers. Early rotary converters would not operate stably unless the frequency of the alternating current they were to convert was low—usually 25 Hz. Compared to higher (60 Hz) frequency current, such low-frequency current was not desirable for users because it would not work with arc lights, it caused incandescent lights to flicker, and it required more expensive motors and transformers.[31] When the final users of electricity received direct current, a utility would generate and transmit 25 Hz alternating current and convert it to direct current before distribution. When the final users were to receive alternating current, 60 Hz generators and transmission systems were used.[32] Motor-generators could convert 25 Hz AC to 60 Hz AC, making it possible for a system designed primarily to serve a DC load also to provide AC for final consumption, but this entailed efficiency losses.[33] Even after technology reduced the comparative benefit that 25 Hz AC had over 60 Hz when converted to DC, generation of lower-frequency AC dominated in large cities where DC had long been used. In 1919 New York, 84.6 percent of generation was at 25 Hz. The corresponding figures for Washington, DC, and Baltimore were 99 percent and 98.5 percent, respectively.[34] Nevertheless, alternating current allowed larger integrated networks. As early as 1897 and 1898 the greater the usage of AC by a utility, the lower the total generating capacity required to serve a given load.[35]

The special censuses of 1902 and 1907 showed that over that five-year time period, generating capacity of all DC generators changed very little while the total capacity of AC generators more than tripled.[36] Although AC generating capacity in 1907 was four and half times that of DC, most of the AC was converted to DC for final use.[37] Not only did former users of direct current continue to receive it after alternating current dominated generation, even some customers in areas receiving utility service for the first time were given direct current.[38] In New York, direct current distribution systems remained in use until 2007.[39]

One of the first urban areas to develop a large network was the city of Chicago. The history of electrification in Chicago has received attention from historians, partly because Chicago in the 1910s and 1920s had a claim to having the best electricity supply in the world and partly because of the individual most associated with Chicago's electrification, Samuel Insull. As a teenager, Insull worked as an assistant to Thomas Edison. He became one of the most important people in Edison's organization. After Edison's involvement with the industry ended, Insull moved to Chicago to assume the presidency of the Edison licensee in that city, one of several electric utilities operating there. Through a combination of political skill, business acumen, a willingness to take risks, and luck, he gained sole control of Chicago's electricity supply. He guided his electric utility to the pinnacle of the industry and he became the most prominent executive in the nation's electric power industry. His rise was followed by an even more precipitous fall, discussed in chapter 5, as business reversals led to bankruptcy and criminal prosecution. He spent much of the end of his career as a reviled symbol of business excesses of the 1920s both before and after a high-publicity trial that resulted in his acquittal of all charges.[40]

A paper presented at a meeting of the Association of Edison Illuminating Companies in 1902 gives a glimpse into early network formation in Chicago. The beginnings of a Chicago network preceded the use of alternating current when four previously separate stations in an area of less than five square miles were tied together with direct current. Although the conductor expense was quite high, and no additional generating capacity was provided, "it resulted in a decided improvement in the character of the service and also made possible some saving in operating expenses." Eventually, rotary converters enabled transmission to be done with alternating current that was converted to direct current for use. This allowed the network to encompass a larger area and provide greater benefits. In five years the network that had had interconnected four generating stations contained fifteen generating stations and substations. The author credited the network with significantly decreasing the cost of service.[41]

Metering and Pricing Electricity

Electricity can neither be easily stored nor resold from one purchaser to another. This makes the economics of electricity production more like that of a service, such as transportation or shipping, than like most manufactured products. An electricity user also cannot purchase electricity at one point in time for use in another. These characteristics give electric utilities the need and the ability to adopt complicated price structures. A utility can charge different customers different prices for the same electricity use, and it can charge the same customer different prices for electricity, depending, for example, on the time in which the electricity was used or on the total amount of electricity used by that customer. Since an electric utility does not have competition in the same way as most producers of manufactured products, at least some customers can be charged prices in excess of the cost of serving those customers; competitors cannot offer those customers lower prices.

The difficulty of storing electricity complicates the cost of producing it. The cost of much of the capital equipment that dominates an electric utility's total costs depends more on the maximum (peak) amount of electricity demanded of the system at any point in time than by the total amount used over a long period.[42] Any increase in the use of electricity during the system peak imposes far more costs on the utility than increases that occur at other times. An increase usage during the peak will require investment in additional capital equipment; increases in usage at other times will not. It has always been difficult to know in advance exactly when peak usage will occur, but factors such as the time of day and the season tended to be good predictors of electricity usage.

Edison measured the usage of the customers of his Pearl Street station in 1883 with an electrolytic meter. The meter actually measured ampere-hours, but since current was provided at a single voltage, this was equivalent to kilowatt-hours, the measure of energy. Edison did not initially price electricity by the kilowatt-hour; instead, he based the price on the amount of light produced by the electricity. He took this position because he wanted improvements in the efficiency of light bulbs not to accrue entirely to the benefit of the customer.[43] Nevertheless, such a method of pricing was nearly equivalent to pricing by energy use. Edison's model for electric utilities, gas-lighting utilities, commonly used meters that measured the quantity of gas used, a close proxy for energy use. However, a price based on energy use was far more appropriate for gas lighting than for electric lighting. Because gas could be stored, the cost of its production was not as sensitive to whether usage occurred on-peak or off-peak. The primary use for electricity was for lighting, and the greatest demand for lighting was in the evening. The prices charged for electricity by the early utilities was high. Although electricity provided a more desirable light, its price was higher than the same amount of light produced by gas. The use of electricity was very uneven; most of it occurred in the evening. Some early utilities did not even bother to provide electricity during the day, and some did not even provide it during evenings when the moon was bright.[44] If electricity was priced by energy use (kilowatt-hours), and nearly all usage occurred at the same time during the evening, the price per kilowatt-hour had to be high to cover the capital equipment costs. That high price would discourage use at other times even though the cost of providing it then would have been very low. Energy-based pricing resulted in inefficient consumption patterns that caused high prices that reinforced those patterns.

When utilities first began providing alternating current, there was no way of metering it. Payments to the utility depended on the amount of electricity-using equipment connected to the utility's circuit or on a physical characteristic, such as number of rooms of the building provided with electricity. A common method was to base the cost of electricity on the number of bulbs of some specific candlepower a customer had attached to the system. Unlike Edison's energy-based pricing system, a charge per bulb encouraged customers to use their lights throughout the day. This would likely have resulted in a lower average cost to the utility of providing lighting. On the other hand, a per-bulb charge would have discouraged customers from installing more than the absolute minimum number of bulbs and made it necessary to monitor bulb use to prevent the surreptitious attachment of unreported bulbs. Customers were supposed to obtain their bulbs only from their utility. An early utility providing alternating current could avoid this problem by using a rate that charged a flat amount per house or based the charge on the number of bedrooms. This avoided the monitoring problems of a per-bulb charge but provided a large barrier for potential customers who wanted to determine whether the new lighting system warranted paying for electricity. Of the 952 companies surveyed by the Commissioner of Labor in 1897–1898, 522 were providing "unmetered incandescent service."[45] The report attempted to provide data on the prices charged by different utilities but found it impossible to do so in a way that was comparable across utilities: "The prices for incandescent lamps are, therefore, entered just as they were contracted for—that is, per year, per month, per week, per kilowatt hour, per ampere hour, etc." The units provided include "Kw.hr.," "Amp.hr., " "Lamp per mo., " "Lamp per wk., " "12 lamps per mo.," "House per mo.," "Lamp per night," and "Lamp hr."[46] Some utilities offered electricity for motors at a lower rate than electricity for lights. Since motors were more likely to be used during the day than were lights, the additional cost to a utility of providing electricity for a motor was likely much less than that of providing it to additional lights.

Overall economic efficiency occurs when the price of a product equals its marginal cost—the addition to total cost of producing and selling an additional unit. Users then equate the benefit of the last unit used with the cost of producing it. This will maximize the difference between the total value of the electricity used and the total cost of producing it. Unless there are significant differences in the utility's equipment required to serve different customers, there should be little difference in the marginal costs of serving different customers. The major price difference would differentiate electricity used during or near the system peak with that used when generators or other capital equipment was idle, and this difference would apply to all customers. Without some adjustment, prices based on marginal cost might not have resulted in a utility's total revenue covering its total cost. This situation was particularly likely before 1940. For the utility to be profitable, of course, total revenue must at least cover total cost. Because technological progress was constantly lowering the cost of producing electricity, it is likely that average prices had to exceed marginal costs to cover the utilities' total costs. Still, large differences in the marginal cost of providing electricity during the system peak and at other times justified pricing designed to reflect those differences.

Rather than using prices that varied according to whether electricity was used on or off peak, utilities came to classify customers and charge different prices based on a particular customer's classification. Although differences in the costs of producing and distributing a kilowatt-hour could justify the use of different prices, this was not the only reason for charging different prices. The economics of electricity production meant that charging different prices to different customers, or even different prices to the same customer, could increase utility revenues and profits even when there were no differences in the marginal costs of producing and supplying the electricity. This basis for pricing is termed "price discrimination" by economists but was sometimes referred to in the industry as "value of service" and by skeptics as "charging what the market will bear." The higher profits provided by price discrimination did not necessarily mean those profits were excessive, although they could have been. Higher profits might have simply enabled a utility to avoid losing money.[47] Price discrimination could have enabled an electric utility to serve customers it otherwise would have found unprofitable, and it could even have resulted in lower prices to everyone.[48] The benefits of electricity grew over time and soon became so large as to significantly exceed any of the prices actually charged, but an awareness that others were paying a lower price created dissatisfaction or even anger. To deal with this, utilities appealed to arguments based on cost.

As long as the price charged by a utility for increased service exceeded the additional (marginal) cost of providing that service, the utility's revenue and profits would increase as the use of electricity increased. This, however, required that the lower price apply only to the increased service induced by that lower price. If the lower price also applied to the service that would have been purchased at the higher price, utility profits could have suffered. The objective of price discrimination is to apply lower prices only to that additional service, although this can never be perfectly achieved.

Declining-block price schedules provide an obvious example of a method of charging different prices to the same customer in an effort to charge a lower price only for additional usage. In this type of rate structure, electricity usage is divided up into blocks of fixed amounts of kilowatt-hours per month. Each block is priced below the previous block, but that lower price only applies to the electricity in that block.[49] As long as a customer's use exceeds the amount of the first, highest-priced, block, the cost to a customer of using an additional kilowatt-hour is less than the average cost of all the electricity used. When deciding whether to increase the use of electricity, the relevant price to the customer is the lower price, not the average price, and the customer would likely use more than if all electricity were priced at the average. Unless the total cost is a substantial portion of a customer's income, the amount of electricity used would be close to that which would have been used if all electricity had the lower price, but the total revenue received by the utility would be greater.

When different prices are charged to different customers, profit is increased when the lower prices are charged those for whom the lower price will induce the largest increase in usage (higher price elasticity of demand).[50] The use of customer classes enabled utilities to do this. The major factor determining price sensitivity was the extent to which a customer, or potential customer, had good alternatives to the use of electricity provided by a utility. Price sensitivity (demand elasticity), rather than production costs, became an important factor in the rate schedules used for customers within a particular class. An important example of a rate structure feature that targeted electricity users on price sensitivity rather production costs was the widespread use of the "demand charge" in rate structures for industrial users of electricity. The term "demand" within the utility industry referred to a customer's maximum power usage regardless of total energy use. For example, an electricity user that consumes one kW of power continuously for 100 hours uses 100 kWh of energy but has only one kW of demand. Another user that consumes 100 kW continuously for only one hour uses the same amount of electric energy but has a demand of 100 kW. A demand-charge rate structure bases a customer's bill on both energy use and demand. Under such a structure, the second customer would have higher bill than the first. Those whose use of electricity was steady paid less than those whose use fluctuated.

A demand-charge rate structure superficially mimics the cost to the utility of providing electricity to all its customers. The cost to a utility is higher if maximum (peak) power use increases even if total energy use is unchanged. Demand-charge rate structures do the same for individual customers. Despite the apparent similarity, demand-charge rate structures fail to reflect the utility's costs because those costs are determined by the amount of power the customer uses during the system peak, not the maximum amount of power that may be used at other times. A rate structure that truly reflected the utility's costs would charge a higher price for all uses of electricity during the system peak and lower prices at other times. The extent to which any individual customer's use of electricity fluctuates is irrelevant.[51] Utilities defended their use of demand-charge rate structures by making the false claim that they reflected the cost to the utility of serving each customer. Since the claim was false, why were those rate structures so widely used?

Industrial users had a good alternative to purchasing electricity from a utility: generating it themselves in their own isolated plant.[52] An industrial plant that generated its own electricity faced the same economics as a utility: a significant determinant of the cost of self-generation was the maximum power that plant was required to produce, and that was determined by the plant's peak usage of electricity. A demand-charge rate structure was not based on the cost to a utility of providing electricity. It was based on what it would cost an individual customer to use the alternative source, an isolated plant. The amount of electricity an industrial user would purchase from a utility depended on whether the cost of the utility-provided electricity was more or less than the cost of using an isolated plant. A demand-charge rate structure enabled utilities to charge higher prices to those for whom the alternative was more costly and lower prices to those for whom it was cheaper. It was an effective means of price discrimination.

Those in the electric utility industry understood the difference between the effects of the system peak versus that of the individual customer on the cost of electricity production. It is clear from the historical record that the basis of the demand-charge rate structure was price discrimination, not

production costs.[53] A wide-ranging discussion on the proper basis for pricing electricity occurred within the electric power industry from the late
nineteenth century until state commission regulation was adopted in the
middle of the first decade of the twentieth century. That discussion was international in scope, occurred among both engineers and industry executives, and was a topic of articles in engineering and trade journals and in
discussions at meetings of electric utility trade organizations. There were
advocates of time-of-day rate structures, which would have based rates
more closely on production costs, and meters to implement such rates
were developed before the turn of the twentieth century.[54] The eventual
widespread adoption of demand-charge rate structures provides insight
both into the nature of the business climate faced by the young utility
industry and into one of the effects of state regulation.

About ten years after the founding of the Pearl Street Station, electrical
engineers first began dealing explicitly with the problem of developing a
theoretical analysis of rate structures. At this time, residential users were
utilities' primary customers, and their maximum power use was likely to
occur during the system peak. The earliest advocates of demand-charge
rates had residential, not industrial users, in mind. Some of the most active
proponents of particular rate structures had an economic interest in the
supporting metering technology. These included Arthur Wright, the British inventor of a demand-charge meter who tirelessly promoted his meter
on both sides of the Atlantic, and Samuel Insull, who became a backer of
the Wright meter.[55] The wide-ranging discussion also included other proposed rate structures. In 1902, the discussion within the industry decreased
markedly. Evidence on the actual use of rate structures is very sparse but
suggests that time-of-day rates had little if any use in the United States
and that demand-charge rates were primarily applied to residential use,
seldom to businesses. There appears to have been little uniformity within
the industry on the use of rate structures.

Apparent neglect of the issue of rate structures ended in the next couple
of decades. When states began to adopt rate regulation, the electric power
industry lobbied forcefully for the use of demand-charge rate structures
for industrial users. Their reasoning was quite clear, as expressed by a 1914
report from the NELA's Rate Research Committee to the Association:

> In the case of large customers, the value of the service to the customer clearly
> depends on the amount for which he could make the same service for himself,
> because if the rate asked is notably higher than this amount, the customer may
> put in his own plant. The value of the service to the customer depends on what

it would cost him to make it himself, and this cost clearly depends in part on the size of plant that he would need. The size of plant that he would need is determined by his maximum demand and necessary reserve. . . .

The demand is at least a rough measure of this cost, and is therefore a test of the value to the buyer.[56]

As early as 1909, regulatory commissions accepted demand-charge rate structures as instruments of price discrimination, as did textbooks on electric rates as late as 1933.[57] The importance of this type of price discrimination stemmed from the fact that isolated plants were a very serious source of competition for electric utilities at the beginning of the industry and for many decades after.

The Challenge from Isolated Plants

The operators of isolated plants were desirable customers to the early equipment manufacturers because they were in a better position to pay cash for their equipment than were the capital-starved utilities. For a factory that wished to electrify, self-generation was often the best, if not the only, option. The blow this caused a young utility was softened by the provision present in the early franchise agreements that the utility would receive a commission for every isolated plant sold in its territory. Even after the electric utility industry became better established, isolated plants were an attractive alternative to electric utilities for a major portion of all electricity use in the nation.

The 1902 special census omitted isolated plants from its data collection but had this to say about their importance:

> In fact, no statistics of isolated plants are included in this report, which, to that extent, therefore, falls short of embracing the entire electric light and power industry of the United States. Many of these isolated plants are of a very extensive and important character, being supplied with the most improved apparatus and giving facilities equal to those furnished to populous communities. It is estimated that there are 50,000 of these plants, and that they consume at least half the product in some lines of electric apparatus.[58]

Despite this statement, that census provides some data on the production of electricity from nonutility plants, particularly isolated plants used in larger industrial facilities, mines, and electric railways. The types of isolated plants for which data was absent included those in institutions, hotels, apartment houses, office buildings, and amusement parks. Even though available data underestimate the amount of electricity produced by nonutility plants, they clearly show its importance (figure 2.3). At least until 1912 isolated plants actually produced more electricity than the entire utility industry. By the beginning of World War II, they still accounted for at least one-fifth of total generation. Data on the source of electricity used to power motors used for manufacturing are even starker (figure 2.4). For the years 1904 through 1914, when state regulation was becoming common, isolated plants were a much more important source of electricity for this use than were utilities. Industrial customers were desirable customers, but they were not easy to get, and utilities faced significant competition for their business. The problems of competing with isolated plants were a frequent topic of discussion at meetings of utility trade groups, and direct advertising was used to persuade the switch from isolated plants to utility service (figure 2.5).

Isolated plants had a number of economic advantages over utilities. As indicated by the quote from the 1902 census report, the optimum size of generating equipment was initially low enough that many isolated plants were equipped with exactly the same equipment that would be used by a

FIGURE 2.3. Proportion of electricity made available in the United States from nonutility sources

Source: Cintrón and Edison Electric Institute 1995, 53

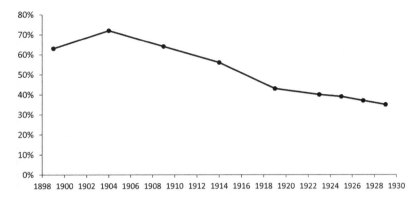

FIGURE 2.4. Percentage of electricity used to power motors in manufacturing generated by the manufacturer

Source: US Bureau of the Census 1932, 112

large central-station utility. A factory using an isolated plant avoided the costs associated with transmission, distribution, metering, and billing. In addition, the generation of electricity required the production of steam, and steam was useful in many industrial processes. Once steam had passed through a steam engine or turbine for the generation of electricity, it could be used for space heating or for an industrial process. These multiple uses for steam enabled a factory to use fuel more efficiently than could a utility. At least one utility tried to blunt this advantage by offering subsidized steam to its industrial customers.[59]

Central-station utilities also had important advantages over isolated plants. Technological improvements in generators tended to increase the size at which a generator achieved maximum efficiency. Eventually, few factories needed enough electricity to warrant such a large generator. As long as a factory's peak occurred at a different time than the system peak, a utility would have required less additional generating capacity to serve the factory than the factory would have needed to serve itself. In addition, the additional backup capacity required by a utility was less than that required by an isolated plant to achieve the same level of reliability.

The slow decline in the importance of isolated plants over time is evidence of the increasing relative advantages enjoyed by utility-generated power. The change was not sudden, however, and the issue of whether an industrial user of electricity was better off purchasing electricity from a utility or self-generating it was contentious for many years. Consulting

Electric Light From Central Station Is Cheapest

The bill at right shows $3.80 as the net cost for light (40 kilowatt hours) supplied by *Central Station*. Bill at left shows $6.40 as what net cost for same number of kilowatt hours would be in an office building maintaining its own power plant. The cost of electric light supplied to tenants by office buildings equipped with private power plants, is generally much higher than the cost in buildings whose tenants are supplied electric light from Central Station. In some instances the rate is as high as 16c per kilowatt, while Central Station rate is only 10c per kilowatt.

The Truly Modern Building Uses Central Station Current

Only a few years ago these office buildings having their own power plants for elevator service and electric lighting were considered distinctively modern. But today the very converse is true, and such buildings are now regarded as among the less desirable. One of the more distinguishing features of the truly modern office building is the use of power and light from Central Station.

Visitors are always welcome at Electric Shop, our show rooms, corner Michigan and Jackson Boulevards, where electric light bills may now be paid, and arrangements be made for the exchange of used-up lamps.

Commonwealth Edison Company
Telephone Randolph 1280 — 139 Adams Street

n D. McJenkin Advertising Agency

FIGURE 2.5. Commonwealth Edison advertisement on the superiority of centrally generated electricity over isolated plants
Digitally enhanced to improve clarity.

Source: Commonwealth Edison 1909, 12

engineers made a business out of analyzing a factory's requirements for electricity and advising its owners as to whether they were better off purchasing or producing electricity. The topic frequently arose in meetings of professional societies and in the pages of professional and trade journals.[60] For many factories the decision they faced was whether to buy any electricity from a utility or to generate it all themselves. If the costs to the factory of the two supplies were close, small changes in the prices charged by a utility could have had a big effect on the amount of electricity sold to factories by the utility. It was in the interest of utilities to target lower prices to those for whom isolated plants were particularly attractive, those whose individual peaks were small compared to their average use of electricity. The demand charge enabled just such a targeting.

The Adoption of State Commission Rate Regulation

E lectric utilities never operated within a free competitive market; they were always subject to government control. Before the creation of state regulatory commissions, that control was exerted by municipal governments. That system led to serious problems, resulting in a movement for reform and a new system of state commission regulation that included control over prices. Such regulation had originally been tried with railroads, but it was ineffective until it began to be used with utilities during the first two decades of the twentieth century. State regulation significantly affected both the structure and behavior of the industry. As regulation evolved, it retained much of its original character and now dominates the US electric power industry in most states today.[1]

State (and, eventually, federal) regulation sought to maintain the benefits of monopoly supply while protecting utility customers from exploitation enabled by monopoly power. Controversy quickly developed over whether this form of regulation effectively protected consumers or operated primarily in the interests of the utilities. There was, of course, considerable variation among states in the competence, objectives, and honesty of state regulators. Nevertheless, even if the aspects of state regulation designed to protect users worked perfectly, it created a flawed structure with characteristics that discouraged the most efficient production and use of electricity. Identifying those flaws is not difficult; designing ways of eliminating them is. State regulation solved the most serious problem that had faced the industry, and it enabled the industry to grow rapidly and become the envy of the world. The industry structure created by state regulation had negative aspects, as have all structures within which the industry has operated. The design of an ideal structure was, and remains, elusive.

Before State Utility Regulation: The Franchise System

In order to provide service, an electric utility (as well as other municipal utilities such as gas and transit) had to operate a distribution system on land it did not own. The only practical alternative was to use the public streets or rights-of-way. Before the institution of state regulatory commissions, this required that a utility receive a franchise from the municipality. A franchise was a legal contract between the city and the utility that could include any conditions accepted by both sides. When utility service was a novelty, municipalities usually demanded few restrictions in order to encourage development, but eventually some restrictions became common. Very often, these prevented an increase in the prices charged the municipal government for its electricity use, such as in street lighting or in public buildings. Sometimes the franchise set detailed rate schedules for all the utility's customers, or granted the city the right to intervene later in the case of rates judged "excessive." Some state laws dictated certain conditions. Kansas at one time required a utility to turn over all profits above a certain amount to the city government. California exempted electric utilities from needing to acquire franchises but gave explicit powers to city governments over utilities, including the authority to set rates.[2] These conditions were often hard to enforce.

Effectively using franchises to regulate a utility was difficult. A city could establish a maximum rate, but this was liable not to be binding as technological improvements reduced the cost of electricity. If the rates were binding, the utility could respond by reducing its quality of service and, therefore, its costs. Franchises were awarded for a specified period, and that created additional problems. A short term was advantageous to the municipality because franchise renewal was the best, or only, time to modify or set new conditions. However, the possibility that a utility might lose its franchise or be forced to accept onerous new conditions discouraged the construction of the facilities required to generate and distribute electricity and made the franchise unattractive. Because of this issue with investment, franchise terms tended to be quite lengthy—twenty-five to fifty years was common, and some had no expiration.[3] Even long-lived franchises caused the same problem with investment when that life was near its end. Occasionally franchises allowed for some type of renegotiation before then, and some franchises included provisions for the city possibly to buy the utility before the end of its franchise with the price determined by arbitration.[4]

As consolidation occurred within the industry, firms that acquired others would acquire their franchises as well. A single electric utility might then be operating under the authority of multiple franchises. In 1908, for example, Chicago's Commonwealth Edison held twenty-three different franchises.[5] The value of any single franchise to a company other than the one holding them all was likely to be low, and the different franchises were likely to expire at different times. This reduced the power of a city to impose new requirements when any single franchise was renewed.

After 1900, franchises increasingly gave municipal governments power to regulate rates.[6] To be able to withstand court challenge, a municipal rate-setting agency needed to be able to undertake an investigation into the utility's cost of providing service, the value of its investment, and the level necessary for a rate of return to be "fair." This would have required considerable resources. In some franchises, disagreement between the municipality and the utility over rates were to be resolved through arbitration, which would also have required resources.[7]

Initially, municipalities granting special franchises often encouraged competition among electric power companies by granting several franchises for the same territory. Some franchises did grant protection from competition, but that protection was often only for a brief time—the first year or two of a much longer franchise.[8] Users of electric power saw no particular advantage in a monopolist serving them. In 1881, Denver's city council passed a resolution essentially promising a citywide franchise to all comers.[9] Sometimes a city awarded a franchise for an area already served by a utility for the express purpose of forcing that utility to lower its rates or improve its service.[10] Competition should work this way, but the special economic characteristics of electric utilities often caused competition or the threat of competition to lead to an unfortunate outcome. Denver provided an example.

In 1883, the Colorado Edison Electric Light Company received a citywide franchise and began operation in Denver. Four years later another franchise, also covering the entire city, including the territory of the first utility, was given to the Denver Light, Heat and Power Company. The two companies soon merged, and the combined company merged with Denver's gas utility, forming Denver Gas and Electric. Gas then competed with electricity by offering a viable alternative to electricity for artificial lighting, and the merger eliminated that competition. By 1901, this company was charging $120 and $150 per year for arc lights to the city government and private merchants, respectively, and residential users were charged 15¢ per kilowatt-hour ($4 in 2014 dollars). In 1900, the city awarded a new franchise for the same territory to a different company, the Lacombe Electric Company, and a modified franchise was granted the next year. This company offered arc lights to the city government at $90 per year, and it cut the price to residential users by two-thirds. A rate war ensued, and the price charged residential users by Denver Gas and Electric fell to 2½¢ per kilowatt-hour. This bankrupted the Lacombe Company, and it sold out to Denver Gas and Electric. The restored monopoly provider inherited numerous contracts to sell residential electricity for 2½¢ per kWh to customers of the former Denver Gas and Electric and 5¢ per kWh to those of the former Lacombe. The new monopolist put itself into receivership with its president appointed receiver. It then managed to get a court order nullifying all the low-priced contracts whereupon it returned the price of a kilowatt-hour to 15¢.[11] Rate wars could drive prices to a level that only covered operating costs, failing to cover the costs of fixed capital equipment of any of the competitors. Entering into competition with an existing utility was very risky. By the middle of the first decade of the twentieth century, there were few cities with a credible potential competitor to the existing electric utility.

Corruption and the Franchise System

With little threat of competition, a franchise became valuable. Facing only the relatively ineffective restraints of the franchise, a utility could exercise monopoly power and enjoy large profits. The possibility of monopoly profits (monopoly rents) created a motivation for corrupt behavior.[12] Utility operations required large amounts of expensive transaction-specific capital equipment (explained in the introduction). These costs created a situation exposing utilities to a special risk that encouraged corruption. In the long run, the revenue a utility received had to repay investors whose funds covered those costs. If revenue did not fully cover those costs, the utility would continue to operate as long as the revenue at least covered bare operating costs. Worn-out equipment would not be replaced, and operating expenses, including equipment maintenance, would be reduced. Service would deteriorate and, unless the situation improved, eventually enough of the capital equipment would wear out that service would cease. However, this would be a gradual process. If the inability of revenue fully to compensate investors became widespread, utilities would have difficulty raising funds needed for new equipment.

The difference between the amount of revenue required for the utility to cover fully all costs, including payments to investors, and the minimum amount at which it would maintain operations in the short run is termed "quasi-rent."[13] The possibility that corrupt city officials could extort quasi-rent, as well as any monopoly profits, resulted in widespread corrupt behavior by both utilities and city officials. The conditions of franchises could enable extortion by, for example, enabling the threat of competition. The extortion of a portion of a utility's quasi-rents by corrupt officials would not have been immediately apparent to the utility's customers, and the subsequent deterioration in service would likely have been blamed on the utility, not the corrupt officials. Utilities had an incentive to participate in corrupt behavior by, for example, bribing officials not to take actions that threatened quasi-rents and monopoly profits. The protection of quasi-rent was in the (long-run) interest of the utility's customers, and the protection of monopoly profits was not, but it is impossible to distinguish behavior protecting one from that protecting the other.[14]

Most of the value of an electric utility's capital equipment (above scrap value) determined the size of quasi-rents. By 1909, the value of capital in electric utilities had reached about $1.8 billion 1929 dollars, which was about 4.6 percent of the value of capital in all manufacturing and 8 percent of that in railroads.[15] Investment costs, however, were relatively more important to electric utilities in the early twentieth century than for other industries. Figure 3.1 shows the value of capital required to produce $1 worth of output for selected utilities and industries. Until 1929, electric utilities were the most capital intensive. Prior to 1914, by which time most states had created commissions to regulate electric utilities, it took over twice as much capital to produce $1 worth of output in electric utilities as in railroads and over ten times as much capital to produce the same value of output as the average for all manufacturing. Even more dramatic was the importance of annual gross investment to electric utilities compared to other utilities shown in figure 3.2. Until 1915, the industry was, each year, spending more on capital than it received in total revenue. The value of monopoly profits can only be a portion of the value of output and is probably a relatively small proportion. Given the size of capital investment compared to revenue, it is reasonable to conclude that the size of quasi-rents was far larger than that of monopoly profits and thus the larger inducement to corrupt behavior.

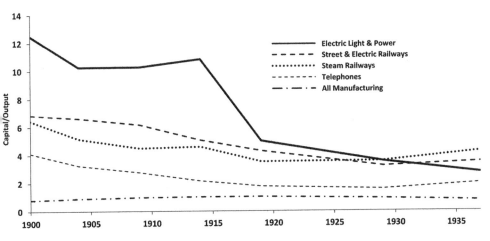

FIGURE 3.1. Capital/output ratios for selected industries

Sources: Creamer, Dobrovolsky, and Borenstein 1960, 265–67; Ulmer 1960, 405–6, 472–73, 476–77, 482–83

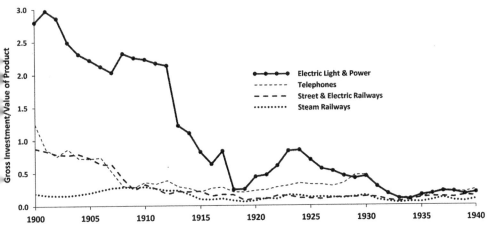

FIGURE 3.2. Ratio of annual gross investment to revenue for railways and selected utilities

Source: Ulmer 1960, 256–57, 320–21, 359–61, 374–75, 405–6, 472–73, 464–65, 476–77, 486–87

On the other hand, the ability of the industry to continue such investments required some degree of confidence among investors that quasi-rent would not be extorted.

Historical evidence confirms that municipal corruption involving utilities and franchises was common. In many cases, the record of corruption involved gas, water, or traction utilities, which were older than electric utilities and larger at the turn of the century. The muckraker Lincoln Steffens cited corruption involving franchises in St. Louis, Philadelphia, Pittsburgh, and Chicago.[16] San Francisco had a dramatic trial starting in 1906 involving graft, franchises, and "Boss" Ruef.[17] In Chicago, corrupt aldermen created new franchises for the sole purpose of extorting existing electric-utility franchise holders by threatening competition.[18] All of these are cities for which the historical record is particularly rich; similar corruption was undoubtedly occurring in many, if not most, other US cities where the record is sparser. The corrupt relationships involving municipal government politics and business interests became a matter of concern to those in the Progressive movement.[19] They recognized that the interaction between private utilities and municipal franchises was a major source of this corruption:

> The chief sources of corruption in American cities are necessarily public contracts and the granting of special privileges or exemptions. . . . the most serious scandals have almost always been connected with or centered around special privileges, and chiefly public utilities. Those who seek special privileges have been most likely to offer illegal inducements to the public official who has the power to grant these privileges. . . . They create and maintain a continuing lobby—always defensive and frequently offensive—because they have a constant incentive through their necessary relations with the public authorities.[20]

Many Progressives regarded the municipal ownership of utilities as a solution to the corruption problem. Municipal ownership may seem an unlikely solution since such an arrangement certainly offered opportunities for such corrupt behavior as supplier kickbacks, employment as reward for political favors, and so on. However, municipal ownership would probably have solved the specific problem of corruption involving quasi-rents. Accountability for poor service would then clearly lie with city officials. They might hesitate to take an action that could jeopardize the ability of the utility to raise investments funds, which could even affect the ability of the city to borrow money for other purposes. The need for transaction-specific capital is generally an inducement to vertical integration. Isolated plants were clearly an example of such vertical integration. This was not possible for most of the customers of an electric utility, whose use could not justify separate generators. Municipal ownership was the closest feasible approximation.[21]

Figures 2.1 and 2.2 show that the proportion of electric utilities munici-

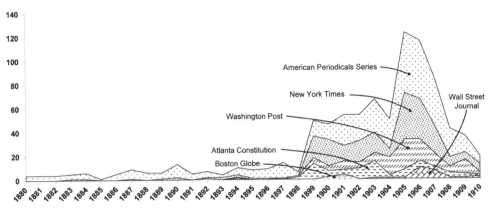

FIGURE 3.3. Annual numbers of articles in the popular press on municipal vs. private ownership of electric utilities

Source: See note 22

pally owned increased, although their share of total generation or number
of customers did not. The privately owned industry, nevertheless, viewed
municipal ownership as a threat and fought vigorously against it. In some
areas, municipal ownership did become significant, and the fear by the privately owned sector of the industry that this might spread to other areas was
not without basis. In 1912, municipally owned utilities had more than half
the total utility generating capacity in two states (Washington and Florida).
For all the years covered by the special censuses of electric utilities (1907
through 1937), municipal utilities were responsible for at least 20 percent
of generating capacity in five or more states. The perennial controversy
over whether it was better for utilities to be owned by government or by
private companies heated up at the turn of the century. Figure 3.3 shows
the number of articles on the issue of private versus government ownership
of utilities that appeared in five major newspapers and in a collection of
periodicals that started publication before 1900.[22] Interest in the topic was
clearly growing up to the point that states began establishing utility regulatory commissions.

The Origin of Rate-Setting Regulatory Commissions

In the early decades of the nineteenth century, state legislatures often
set prices for both private and public businesses, and statutory controls

on the rates early railroads could charge were common.[23] A new railroad required a charter of incorporation from a state government to begin operations. Sometimes these imposed conditions on the railroads, including ones designed to restrain profits and others that provided for the eventual transfer of the railroad to state ownership.[24] Railroads were common carriers, offering service to the general public. As such, they were required by common law to serve all persons under reasonable terms without undue favor and to charge rates that were reasonable and just.[25]

State legislatures had long frequently created investigating committees, including permanent investigating commissions when warranted. Such commissions were created for various industries, including railroads, but they were concerned with safety, not rates.[26] In 1869, Massachusetts created the first railway commission, the model for "weak" commissions later created in other states. It had the authority to examine the financial and physical conditions of railroads, and it could determine, but not impose, rates that it regarded as just and reasonable. Although it could make recommendations to the railroads, and propose legislation, it had no power to compel action. Its ability to influence public opinion and affect actions of the legislature and attorney general, however, did give it considerable influence.[27]

Effective government control over railroad rates was difficult. Like electric utilities, the services of a railroad could easily be neither stored nor resold, and this made price discrimination possible. As with electric utilities, the importance of the joint costs of capital equipment and operations that simultaneously served all customers both increased the extent to which prices could vary among groups and enabled railroads to assert that price differences were entirely based on the differences in the costs of different services. However, profit-maximizing price discrimination based rate differences on the degree to which price changes affected the amount of a service that could be sold. For railroads, this sometimes led to price differences that could not conceivably arise from cost differences. A famous example was the long-haul short-haul situation. This existed when there were several railways offering passage between two large cities, but each railway took a different route between them. Traffic between the two cities was sensitive to the price charged because it was easy to shift that business to a railroad offering a lower price. However, a railroad did not face competition for traffic involving intermediate locations on the route it took, and the amount of such traffic was relatively insensitive to the price charged. Profit-maximizing price discrimination led to lower prices being charged for carriage between the two large cities than between any two points involving an intermediate point even though it was inconceivable that the longer distance was less costly to the railroad than the shorter. Large shippers who had freight that traveled along competitive routes might demand and get discounts or rebates (sometimes secret) for the freight they sent to these intermediate locations. This could give those shippers an enormous advantage over smaller competitors located at the intermediate points.[28] The lower rates that obviously could not be justified by costs seemed unfair, and the establishment of maximum rates would not satisfy those concerns. Finer control over rates would be needed, and that required a commission able to make the needed investigations.

Beginning in Illinois in 1873, developments in some states led to the model of what would become utilities commissions. Laws were passed creating new railroad commissions that had the power to set railroad rates. They could investigate compliance with their orders and initiate prosecution against a railroad. Their decisions could be appealed to the courts, but the laws required such an appeal to bear the heavy burden of proof of showing that the imposed rates did not meet the common-law requirement to be reasonable and just.

Opponents of the new laws claimed they violated the Fourteenth Amendment of the US Constitution. That amendment, among other things, forbade a state from depriving "any person of life, liberty, or property, without due process of law," extending to state governments a restriction that the Fifth Amendment already applied to the federal government. It gave new powers to federal courts to review state legislation. Its original intent was the protection of the rights of freed slaves, but it proved useful to businesses seeking to limit the power of states over their operations. Eight cases challenging the constitutionality of these laws made their way to the Supreme Court, which chose one, *Munn v. Illinois*, as the primary precedent for all of the cases and announced its opinion on March 1, 1877.[29] That decision addressed two Constitutional issues: did the new laws' requirements constitute a "taking" forbidden by the Fourteenth Amendment, and did a state law affecting a business involved in interstate commerce violate the first article of the Constitution, which gave Congress the power to regulate such commerce?

The decision was a forceful endorsement of the power of the state to set rates charged by some private businesses, those that were "affected with a public interest," to ensure compliance with the common-law requirement to be reasonable and just. The ruling did not explain what made a business "affected with a public interest," but common carriers were clearly in that class. Businesses were entitled to "reasonable compensation," but the ruling held this was a legislative, not a judicial, matter. "For protection against abuses by legislatures the people must resort to the polls, not to the courts." The ruling also held that in the absence of action by Congress, states could exercise their powers on businesses involved in interstate commerce. Although the ruling was never formally overturned, many of its provisions were effectively reversed. The power of states to regulate interstate rates was taken away in an 1886 decision.[30] In 1890, the Court effectively reversed itself on the role of the judiciary in determining whether rates were reasonable. "The question of the reasonableness of a rate of charge for the transportation by a railroad company . . . is eminently a question for judicial investigation." The decision did not make clear what that investigation should encompass.[31]

The decision that most profoundly affected state utility regulation came in 1898 in *Smyth v. Ames*.[32] That decision established criteria that rates had to meet to provide reasonable compensation, but the criteria were so vague and contradictory that they crippled railroad rate regulation for decades. Utilities commissions were eventually successful in establishing a methodology that met those criteria, but that methodology brought to utility regulation a number of negative characteristics.

The case involved several railroads in Nebraska whose stockholders argued that a requirement that maximum rates be reduced by 29.5 percent was unreasonable and violated their Fourteenth Amendment rights. Although the law required that challenges to it originate in state courts, the plaintiffs immediately went to the federal courts, establishing the precedent that those challenging regulation were free to choose the court system they wished.

The decision of the court was based on appallingly bad economic reasoning and was so biased in favor of the railroads that it exceeded even the arguments of the railroad's lawyers.[33] To meet the requirement that rates provide "reasonable compensation," the decision held that the revenue provided must cover the company's operating costs and a fair return on the fair value of its property: ". . . [T]he basis of all calculations as to the reasonableness of rates to be charged by a corporation . . . must be the fair value of the property being used by it for the convenience of the public."[34] In its decision, the court avoided determining the value of the railroads' property, because it concluded that the mandated rates would not even cover their operating costs. Instead, it provided vague guidance on the determination of that value that was less than helpful:

... in order to ascertain that value, the original cost of construction, the amount expended in permanent improvements, the amount and market value of its bonds and stock, the present as compared with the original cost of construction, the probable earning capacity of the property under particular rates prescribed by statute, and the sum required to meet operating expenses, are all matters for consideration, and are to be given such weight as may be just and right in each case. We do not say that there may not be other matters to be regarded in estimating the value of the property. What the company is entitled to ask is a fair return upon the value of that which it employs for the public convenience. On the other hand, what the public is entitled to demand is that no more be exacted from it for the use of a public highway than the services rendered by it are reasonably worth.

Each of these methods of valuing property would give a different result. Particularly problematic was using the value of a company's stock to value its property. The market value of stock is determined by investors' expectations about the company's future profitability. Higher rates would increase the firm's profitability and the value of its stocks. If that higher value were then used to justify even higher rates, a spiral of ever-increasing rates would be supported.

An effect of the *Smyth* decision was that the method used to value a utility's property became a constant matter of public and judicial controversy, especially when there was a change in which method resulted in the highest evaluation. The use of "original" cost of construction led to a valuation based on what the utility paid for the capital equipment when it was first acquired minus depreciation. This was also called the "prudent investment" approach. "Present" cost of construction came to be known as "reproduction cost." This required a determination of the costs under current conditions of duplicating the utility's capital equipment less depreciation. Particularly when regulation was first applied, but later as well, determining the original price a utility paid for its equipment was occasionally very difficult. Sometimes there were no good financial records to support this determination, particularly if a utility had changed hands. Reproduction cost, however, always involved speculation when determining the current cost of duplicating obsolete equipment constructed using methods no longer used. Between 1900 and the Great Depression, because construction costs were increasing, the reproduction cost approach justified higher rates, and it became the most widely used. Another contentious issue was whether "going value" should be included in the value of a utility's capital.

Going value was a measure of the difference between the total value of a business to another owner and the value of just its physical capital. Since it was partly determined by rates, using it to determine rates created the same circularity as using stock prices. Nevertheless, its inclusion received support from regulatory commissions and courts. Other contentious issues further complicated the determination of the value of a utility's capital equipment.[35]

Smyth established other characteristics of the methodology commissions eventually used to determine rates. The requirement that rates cover operating costs and a return on capital equipment had to be applied to each class of customer. This meant customers had to be divided into classes and joint operating and capital costs had to be divided somehow among all the classes. As with railroads, there was no objective way of doing this. The only requirement was that the divisions be "reasonable." Many different methods for doing this arose, each of which allocated those costs differently and each of which was arguably reasonable. This provided a mechanism for utilities and commissions to implement price discrimination. It also created an opportunity for controversy over which method was fair.

Determining whether new rates were reasonable required determining the total revenue they would provide the utility. Following *Smyth*, this was done by assuming the new rates would have no effect of the amount of electricity customers used. During most of the period covered by this book, electricity prices were falling and consumption was increasing. The assumption that usage would not change generally underestimated future revenue, supporting higher rates.

The situation created by *Smyth v. Ames* continued until it was overturned by the Supreme Court in the Hope Natural Gas case in 1944. The new decision took the issue of deciding the correct methodology away from the courts and strengthened the power of regulatory commissions. If a decision was challenged, courts were directed to presume that whatever method was used by a commission to value a utility's property was valid. Instead, a successful challenge had to show that the consequences to the utility were unjust and unreasonable. Such a challenge carried a "heavy burden."[36]

State Commissions, the Federal Government, and Regulatory Effectiveness

The 1886 decision denying state commissions the power to regulate interstate rail traffic put 75 percent of railroad business beyond any regulation.

Congress had long considered the interstate regulation of railroads, but the 1886 decision spurred it to action. In 1887, Congress created the Interstate Commerce Commission (ICC), but ten years later, the Supreme Court ruled that it lacked the power to set rates. Once again, most of the business of railroads was completely unregulated.[37] In 1906, when state commissions were first being established to regulate utility rates, Congress passed the Hepburn Act giving the ICC clear authority to regulate rates.[38]

Between 1897 and 1906, only the few "strong" regulatory commissions retained any power over a portion of railroad rates. Railroads, however, became adept at using the courts to block that power.[39] As soon as a state commission issued a rate order, the railroad would appeal to the courts, and the court's task was to balance the Constitutional right of the railroad against the common-law right of its customers. The balance tended to tilt more towards the railroads than the regulatory commissions.

Other techniques impeded efforts of regulatory commissions. Once a commission had decided on rates, railroads sought, and often received, temporary injunctions on the implementation of those rates until the court could review them. The injunctions blocked the application of the rates to all railroads, not just those making the appeal. Repeated appeals could extend the injunctions, often for years, by which time the rates were no longer relevant. Litigation was used to exhaust the resources of a regulatory commission. Although commissions offered full hearings, railroads would give them only a perfunctory effort and then appeal to the courts. Relevant information was commonly withheld from commissions only to be produced in courts that then overturned commission decisions. During the ten years when the first ICC tried to regulate interstate rates, railroads subjected it to the same treatment.[40] Effective regulation could not happen until reform altered the relationship between regulatory commissions and the courts. The necessary reform eventually occurred when state regulation of electric utility rates began.

In 1905, the Wisconsin legislature enacted a law that created the first new "strong" railway commission in almost two decades. In the same year New York created a utilities commission with power over electric rates. The Wisconsin law explicitly addressed the problem of judicial review. Instead of being able to appeal an order of the commission, a railroad would have to commence an action naming the commission as a defendant and seeking to have its order vacated. The burden of proof lay on the railroad to "show by clear and satisfactory evidence that the order of the commission complained of is unlawful, or unreasonable, as the case may be."[41] The law directed courts to give these suits precedence over all over civil cases and specified that both parties prepare for trial within ten days. A court injunction blocking implementation of a rate required prior notification to the commission and a court hearing before the injunction became effective. If a railroad presented evidence to a court not previously given the commission, the court was to suspend its proceedings for fifteen days while the commission considered the new information and decided whether to rescind or modify its order. If the commission rescinded its order, the court was required to dismiss the suit. If the commission modified its order, the court could only consider the new order in its decision. When Congress debated the Hepburn Act in 1906, the issue of judicial review was the most contentious aspect of the debate, and Congress incorporated the Wisconsin provisions for appeals to the federal courts.[42] The following year Wisconsin extended the power of the regulatory commission to cover utility rates.

By 1912, when state utility commissions were well established, some, but not all, state court systems were showing deference to the rulings of regulatory commissions.[43] The federal judiciary, especially the Supreme Court, was showing even greater reluctance than state courts to overturn the actions of administrative bodies including state and federal regulatory commissions. A 1909 Supreme Court decision held that the standard to overturn an act of the New York legislature and a decision of its regulatory commission required ". . . showing beyond any just or fair doubt that the acts of the legislature of the State of New York are in fact confiscatory."[44] States learned how to design laws creating regulatory commissions that could effectively function and still respect the rights of regulated firms defined by the *Smyth* decision. The regulatory processes adopted by these commissions required them to value the utility's property and determine its level of operating costs using methods that required the dedication of considerable resources by both the commissions and the regulated utilities.

Effective regulation of railroad rates by state commissions was short-lived. In 1914, the Supreme Court ruled that the power of the Interstate Commerce Commission to regulate interstate rates included the power to overrule a decision of a state commission on a rate for transport wholly within the state.[45] Most state commissions were left with only the power to regulate utilities. This blow to the power of state commissions was not forgotten. A national association of state regulatory commissioners had long existed, and by 1914 it was called the National Association of Railroad and Utilities Commissioners (NARUC). Not until the 1970s was the word "Railroad" removed from the organization's name (and replaced with "Regulatory"). Decades later, when Congress was considering legislation that eventually established federal regulation of interstate electricity rates, state regulatory commissioners became very active in opposing any measure that might have threatened their authority over electric utilities (see chapter 6).

The Adoption of State Commission Regulation of Electric Utilities

In June 1898, Samuel Insull delivered the presidential address at the annual meeting of the National Electric Light Association (NELA), the national trade group for privately owned electric utilities. His address included an assertion of the desirability of a system of government regulation of utility prices designed to restrain profits. The other utility executives were less than enthusiastic. Insull gave no details as to what government or government agency he thought should be responsible for regulating the prices for electricity. His remarks followed the *Smyth v. Ames* decision by three months. There is no evidence that he had the decision in mind, but the position he took was consistent with it and with the discussions about state regulation that occurred later. His arguments included:

- The adoption of rate regulation is, at least in part, a reaction to growing interest in municipal government ownership of electric utilities.
- Competition is an ineffective means of regulating public utilities. This stems from the heavy investment such an enterprise requires. The threat of competition "frightens the investor," thus raising the utility's largest cost—interest payments. Competition inevitably ends with consolidation with the new monopoly saddled with higher costs. Insull did not explicitly mention municipal corruption, but he described the conditions that created the corruption.
- Regulated rates should ensure that a utility receives "cost plus a reasonable profit." This would result in lower interest rates, lower total utility costs, and lower rates for consumers.
- A municipality should have a "right of purchase" over a private utility. This would encourage the private owners of the utility "to do their full duty in their relations to the public." Following such purchase, the municipality would own and operate the utility. The municipality could not go into competition with a privately owned utility prior to purchasing it. "If this is not done, the value

of private property will be destroyed, without just compensation being made therefor [sic], in an attempt to secure a public benefit." The price paid would also fully cover the costs of the private firm's capital equipment handed over to the municipality.[46]

Although not immediately successful in convincing his colleagues, Insull established an NELA committee of like-minded utility executives to study legislative policy on utility regulation.[47] That committee was inactive and eventually disbanded but reestablished in 1905.[48] By then, industry views on the desirability of regulation had changed, partly because of increased concern over municipal ownership.

Several efforts attempted to bring together people on both sides of the ownership issue to examine the merits of both private and government ownership and the role of regulation with private ownership. In 1899, the NELA proposed a joint investigation with the League of American Municipalities at the latter's convention and provided a check for $2,500 to cover half the NELA's share of expenses.[49] The president of the NELA proposed renewing the offer in 1902 and, as a possible alternative, extending a similar offer to the American Economic Society (forerunner of the American Economic Association), whose members were seen as advocates of municipal ownership.[50] In February 1903, the New York Reform Club held a three day "National Convention on Municipal Ownership and Franchises." The convention heard from numerous speakers on both sides of the issue, but there was no apparent move toward consensus. Some of the discussion concerned the British experience, where higher proportions of different types of utilities were municipally owned.[51]

In September 1899, the Chicago Civic Federation formed the National Civic Federation (NCF), consisting of people with diverse views, backgrounds, and geographic locations in order "to provide for a thorough discussion and consideration of questions of national import affecting either the foreign or domestic policy of the United States."[52] The organization consisted of academics and leaders of industry, commerce, and labor. It exemplified the ideals of the Progressive movement, and it developed a reputation for attempting, and sometimes achieving, conciliation in labor disputes.[53] After some previous attempts, in September 1905, the National Civic Federation undertook "to determine impartially and scientifically the relative merits of private and public ownership and operation of public utilities."[54]

The NCF created a 150-member Commission on Public Ownership and

Operation, which appointed a twenty-one-member Committee on Investigation, charged with doing the actual study. The latter committee consisted of equal numbers of people regarded as opposing municipal ownership, favoring municipal ownership, or having no position. All subcommittees maintained the same balance. Expert staffs were hired. The investigation included gas, electric, water, and transit utilities in both the United States and Britain. Most privately owned utilities in the United States refused to allow an examination of their books or a physical valuation of their property, a stance that seriously hampered meaningful comparisons with municipally owned utilities.[55] Nevertheless, the committee ultimately issued a ponderous report in 1907 occupying three volumes and nearly 2,500 pages. The report's first volume, almost 500 pages long, was presented as a "popular exposition" of their findings and was made available in paperback at a price of $1 ($25 in 2014 dollars). The second two volumes included detailed reports from experts on utilities.

Except for utilities that provided water, the committee reached no shared conclusion on whether privately owned or government utilities better served the public. There was disagreement over whether "better" meant lower prices, higher efficiency, greater willingness to provide service to unserved areas, better treatment of employees, or something else. Many of those for whom the issue was relevant had stakes in the answer or had already accepted an ideological stance that predetermined their position. Even if there had been clear agreement on the measures of excellence, the operating circumstances and accounting procedures of the two classes of utilities were different enough to complicate comparison.[56] Each person involved with the study was able to find support from its findings that confirmed the position already taken. The various reports, especially in the first volume, have an unsettled quality, alternating between those explaining that the investigation strongly confirmed the superiority of municipal ownership with those asserting that the investigation confirmed the opposite.

Despite disagreement on the primary objective for the study, there was agreement on some issues. Everyone agreed that government should operate water utilities ("a public utility which concerns the health of the citizens"). They also agreed that municipal ownership should not extend beyond utilities to other enterprises. Echoing Insull's position, all but one member agreed that a municipality should be able "upon popular vote under reasonable regulations" to operate public utilities and to buy out an existing private utility "paying its fair value."[57] Such a provision would operate as an ultimate inducement to a private utility not to abuse its customers and "will tend to render it unnecessary for the public to take over the existing utilities or to acquire new ones." Everyone agreed that utilities could not operate in a competitive market; they needed to operate as monopolies. They agreed that the only alternative to municipal ownership was private ownership under government regulation, that "the question is whether it is better [for government] to regulate or to operate."[58] By this time, that position was not novel.[59]

Although the committee agreed on the need for regulation, there were few details on how this would be done. There was agreement that the regulatory authority should have the power to impose a uniform system of accounts and make public all data concerning finances and both cost and quality of service. Utilities would have to seek prior approval before issuing stocks or bonds, but no criteria for approval was suggested. However, the report was silent about the level of government that should be responsible for regulation, the procedures regulators should use, and whether they should have power over rates.[60]

Despite the prominence of the NCF and the effort that went into the study, its complete failure to bring about any reconciliation between the sharply differing views on proper utility ownership hardly made it a blueprint for action. However, once the report was issued, the business interests in the organization were able to get the federation as a whole to interpret it (incorrectly) as favoring private ownership of utilities (with government regulation) over municipal ownership and to publicize this as the federation position.[61]

The First State Regulation of Electric Utilities

Massachusetts was precocious in establishing a regulatory commission with the power to lower gas rates in 1885, extended to electric rates in 1887.[62] After an eighteen-year hiatus, New York initiated what became a period of increasingly rapid creation of state utility regulatory commissions. In March 1905, a New York legislative committee (the Stevens Committee) under the direction of Charles Evans Hughes, future chief justice of the US Supreme Court, began an investigation into gas and electric utilities. The investigation found apparent overcharges (victimizing customers) and overcapitalization (victimizing investors) by utilities in New York City.[63] The committee proposed seven pieces of legislation, including mandatory ceilings on the price of electricity and gas and the creation of two state regulatory commissions that would continually monitor those utilities.[64] The legislation was passed, and the new commission was created in June 1905. The state already had a railroad commission, but the new utilities commission had substantially more power, including authority over rates, which the railroad commission lacked. During its two-year life, it began work on a uniform system of accounts, refused to approve the issuance of almost half the stock proposed by the utilities, ordered rate reductions, and brought about a number of voluntary rate reductions.[65]

In 1906, Hughes was elected governor of New York and, in 1907, implemented a reorganization of the state's regulatory commissions. In place of three commissions and the office of the gas meter inspector, two commissions were created, both having authority over railroads as well as gas and electric utilities. One of the commissions was responsible only for New York City. It also had authority over urban rapid transit. Its costs (other than the commissioners' salaries) were to be paid by the city, but the city had no authority over its operations. The other commission was responsible for the rest of the state.[66]

The powers the Hepburn Act gave to the Interstate Commerce Commission were the model for those given the new New York State commissions. They had the power to require a uniform system of accounting, to fix rates and standards of service, to compel testimony under oath, to compel the production of books and records, and to enter upon the property of any regulated firm for the purposes of investigation. Although municipal franchises were still required for a utility to operate, so was the permission of a commission. Even municipally owned utilities were required to get this permission. This enabled a commission to prevent competition against any utility in its jurisdiction. A commission could undertake investigation upon its own initiative and was required to do so if requested by certain municipal officials or by a group of utility customers. Rate schedules were required to be publicly available. Utilities were required to file with their commission any proposed rate changes at least thirty days before the changes were to go into effect. The commissions had broad powers to set maximum rates, but utilities were free to charge less. Given their history of secret rebates, however, railroads had to adhere strictly to the commissions' rates. Prior approval was required before a utility could issue stock or debt with a maturity exceeding one year. Each commission had five commissioners appointed by the governor and paid a salary of $15,000 per year (about $375,000 in 2014 dollars), equal the salary

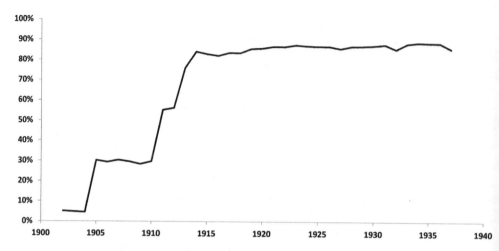

FIGURE 3.4. Percentage US generation in regulated states
Intercensus years estimated by exponential interpolation. Because of census data suppression, years after 1917 omit Delaware, the District of Columbia, and Maryland. Years after 1927 also omit Arizona and New Mexico.

Source: US Bureau of the Census, *Census of Electric Light and Power*, various years.

paid the mayor of New York City and 50 percent more than that paid the governor.[67]

Under the state's Progressive governor, Robert M. Lafollette, in 1905, Wisconsin substantially revised the powers of its railroad commission, including giving it power over rates. In 1907, that power was extended to cover all public utilities.[68] John R. Commons, economist and professor at the University of Wisconsin, was active in drafting the new law. Commons had worked on the National Civic Federation investigation as an expert who favored municipal ownership. Unlike New York's law, where the legislature enacted the bill submitted by the governor without change, considerable negotiation occurred before Wisconsin's legislature acted.[69]

There were some notable differences between the laws in Wisconsin and New York. The Wisconsin law contained far more detail on the procedures involved in determining rates than the New York law. For example, it explicitly required the commission to value the property of each utility and to determine a rate of depreciation for different types of property. The commission's power over rates extended beyond just the setting of maximum rates; it had power over the details of rate structures. All utility franchises were to be of "indeterminate" duration. The law gave any municipality the authority to take over an existing utility at any time after payment of a price determined by the commission. As in New York, the commission did not have to approve rate changes in advance, but changes made by a company had to be filed in advance with the commission (ten days versus thirty in New York). Also like in New York, the commission could initiate the process of changing rates on its own authority. The Wisconsin law contained provisions designed to limit judicial review that were not present in the New York law. Unlike in New York, the Wisconsin commission was not given authority over the issuance of securities.

Other states quickly adopted this new system of regulating utilities (including rates) by a state commission. As shown by figure 3.4, within five years most of the electricity generated in the United States was in regulated states, and by 1914 that percentage was over 80 percent. The new system radically affected the structure and operation of electric utilities. Those states that adopted regulation first tended to be the ones for which the shift away from the system of municipal franchising offered the greatest benefits.[70]

The Effects of State Regulation

State regulation can be expected to have lowered costs for electric utilities by helping to eliminate (or at least reduce) the costs of corruption. In addition, protection against competition improved the willingness of investors to purchase utility stocks and bonds lowering the interest rate the utility had to pay for the funds constantly needed for capital equipment.[71] In addition to reducing interest costs, improved access to capital might have enabled utilities to expand, helping utilities take advantage of scale economies and the lower costs (and perhaps rates) that brought. Offsetting this were a possible reduction in the operational efficiency of regulated utilities because of their protected monopoly status and a reduction in investment resulting from that increased monopoly power.[72] If regulation worked as designed, the revenue allowed a utility would cover its operating costs. If higher efficiency allowed the utility to reduce those costs, after a delay during which the utility would receive all the benefits, the regulatory process would reduce its revenue, shifting the benefits to customers. Although the delay between efficiency improvements and revenue reduction left utilities some incentive to seek higher efficiency, the eventual loss of its benefits reduced them. Wasteful expenditures, such as luxurious offices that did not contribute to efficiency, became more likely under regulation's incentives because those costs could be passed on to customers. Regulatory commissions had the authority to exclude "imprudent" expenditures from the operating costs claimed by a utility, but a regulatory commission could never have the detailed knowledge and information possessed by the management of the utility. Regulation encouraged inefficiencies in utility operations because it reduced the incentives of the utility's decision makers to work hard to eliminate inefficiencies.[73]

The new regulations discouraged efficiency in other ways. The administrative costs of regulation were high. These included not just the costs of the regulatory commission and the costs compliance might impose on a utility. They also included the costs borne by all parties to the regulatory process. Utilities could and would devote considerable resources to making the case that their revenue should be increased. Utility customers would devote resources to making the opposite case. Different groups of customers would likely also devote resources in the effort to reduce the proportion of total revenue they would bear.

The method of regulating rates may have encouraged the use of an inefficient amount of capital equipment. In addition to operating costs, allowable revenue included the value of the utility's property multiplied by a commission-determined rate of return. If that rate exceeded the actual amount required to attract new investment, the difference would have contributed to the company's profits. This plausibly could have resulted in a utility using more than the most efficient amount of capital equipment to increase its profits. This issue received considerable attention within the economics literature with very mixed results.[74] Empirical studies generally found that utilities did use too much capital equipment, but the magnitude and practical importance was unclear. Paradoxically, a theoretical argument was made that a high rate of return would create the opposite incentive—to use too little rather than too much capital equipment.[75] In either case, regulation may have caused utilities to use an inefficient amount of capital equipment.

Regulation introduced a number of other adverse incentives that have received less attention. It eliminated the incentive for utility rates to be structured in ways that promoted the most efficient use of electricity. In a competitive market, the total revenue received by a firm is the sum of the revenue received from each customer. A firm must decide the price to charge each customer, and getting that right determines the firm's profit.[76] Rate regulation inverted this relationship. The central issue of rate regulation was the determination of the utility's total revenue. The allocation of that total revenue among the utility's different customers, and the structure of rates resulting in that allocation, was a secondary matter in the regulatory process and of less concern to the utility than its total revenue. Rates that promoted the most efficient use of electricity were not a priority. Much of the concern of a group of utility customers in the regulatory process would be reducing the proportion of the utility's total profit it bore, and increasing that proportion borne by others. This zero-sum game increased the likelihood that a compromise outcome would involve retaining existing rate structures.

This lack of concern over the structure of rates contrasts sharply with the industry's involvement with the issue, discussed in chapter 2, prior to state commission regulation. This loss of interest is illustrated by the continued use of the demand-charge rate structure long after its economic rationale had passed when rate structures that reflected the large variations in the cost of producing electricity would have better served the needs of electricity users and of society. The anachronistic use of the demand-charge rate structure was a result of the distorted incentives created by regulation.

Regulation may favor a group of electricity users that, compared to other groups, possess either greater political influence, the ability to bring more resources to the regulatory process, or both.[77] This may result in "cross-subsidization," where a more powerful group is able to shift some of the costs of its service to other groups. For example, residential users, the class of users with the most votes, might be able to shift some of their costs to business or industrial groups. A similar mechanism might be present in government-owned utilities, which could be even more sensitive to political pressure than regulators. However, the importance of the cost of operations and capital equipment jointly serving multiple groups (joint costs) makes impossible an objective determination of the total cost of serving any one group. Cross-subsidization would clearly exist if the revenue received from any group of customers were less than the reduction in the costs of running the utility were that group to disappear. However, the many different "reasonable" ways that joint costs could be divided would result in significant differences in relative shares of the utility's cost borne by different groups, none of which would amount to pure cross-subsidization. Pure cross-subsidization clearly occurred, however, because of the failure under regulation to implement rates that reflected the large differences in the costs of generating electricity. Those using electricity during the system peak were subsidized by those using electricity at other times, but this occurred irrespective of customer class. A more tractable concept of cross-subsidization could compare the total revenue of a class of customers with that of an alternative set of prices that was superior by some measure. That would require determining a measure of superiority and using it to define a specific set of prices.

Experience during the last two decades of the twentieth century provides evidence from industries other than electric utilities that regulation is far less effective at bringing low prices than are competitive markets. Until 1978, a federal regulatory agency, the Civil Aeronautics Board, regulated the prices charged by airlines and restricted competition with existing firms, much as state utility commissions did. A 1978 law abolished that agency and freed airlines to choose the routes they would fly and the prices they would charge. By 2000, inflation-adjusted airfares had fallen about 27 percent overall.[78] Abandonment of the regulation of the rates charged by trucking firms had a similar result. In both cases, deregulation profoundly changed the industry's institutional structure and the set of firms in each industry. Although competitive markets are feasible for airlines and trucking, the special economic characteristics of the production and distribution of electricity prevent their full use by electric utilities.

State regulation of electric utilities inhibited the formation of large fully integrated electricity supply networks. The benefits such a network would provide compared to multiple independent networks serving the same area are explained in the introduction. The early utilities both generated and distributed electricity. Within their small service areas, they operated an integrated network. Alternating current and improvements in transmission technology increased the optimal size of integrated supply networks and led to larger utilities. A large-scale fully integrated network could (and very occasionally did) include several utilities under different ownership, but regulation and the importance of transaction-specific capital costs impeded such arrangements. If a fully integrated network spanned the area served by different utilities not under common ownership, transaction-specific capital that served utilities other than the one owning the capital would have been necessary. This is exactly the type of situation that is made unlikely by the presence of risk that would disappear under common ownership. Under regulation, it did become common for utilities to interconnect, permitting occasional ad hoc sales and purchases among them, but this captured only a small portion of the benefits of a network that fully integrated system planning.

The most effective condition for increasing the geographic area of a fully integrated network was for one utility to acquire the service area of an adjacent utility and to consolidate the two. In a competitive market, a large and efficient firm can enter into competition with less efficient firms and reduce their profits to the point that they can be cheaply acquired or driven out of business. Regulation protected utilities against competition and ensured that even inefficient utilities would receive sufficient revenue to continue in business. Protection from competition meant that one utility had to purchase another before it could extend its network. Regulated rates kept up the price such an acquisition would require. Even so, the benefits of large networks brought about considerable consolidation of local utilities after regulation. In 1917, there were 4,224 private utilities, and ten years later, there were 2,137. Average generation by a privately owned utility in 1927 was 5.8 times that of a privately owned utility in 1917.[79] Despite this record, without the impediments created by regulation consolidation would very likely have been even more extensive. Even with the consolidation that did occur, the benefits that even greater consolidation would have brought were considerable (see chapter 4).

State regulation created even greater impediments to the establishment of an integrated network that crossed state lines. It was common for states to require that a utility operating in their state receive its charter from that state, effectively preventing a single utility from operating in another state with the same requirement. Some states explicitly blocked their utilities from participating in integrated networks with out-of-state utilities. In 1924, for example, Maine and South Carolina prohibited export of hydroelectric power outside the state, while Connecticut forbade its import.[80] Agreements between separately owned utilities to create jointly an interstate integrated network were risky. In addition to the general risk, one state's commission might abrogate an agreement to favor its state's users over that of another state. For example, in 1917, a utility in Rhode Island entered into a twenty-year contract to supply electricity to a Massachusetts utility at an agreed-upon price. Seven years later, the Rhode Island commission ordered the contract to be abrogated and the price charged the Massachusetts utility to be raised, which would have reduced the payment required of Rhode Island customers. The case was appealed to the Supreme Court, which, in the 1927 *Attleboro* decision, nullified the Rhode Island's commission action by ruling that a state commission had no authority over interstate transactions.[81] Nevertheless, this illustrated the risk state regulation posed to interstate networks.

State regulation provided the electric utility industry with a new structure unlike any that had existed previously. The new structure was superior to the one that it replaced, but it had flaws, and it did not quiet the industry's political controversy. Instead, it created a new arena for that controversy over whether regulation successfully protected electricity users from exploitation by utilities now firmly insulated from competition. Some of the incentives regulation created for utility decision makers discouraged the most efficient operation of utilities. It allowed the structure of electricity rates to be neglected, encouraging the inefficient use of electricity. It reduced the benefits offered by large fully integrated networks. In the decades following the establishment of state regulation, holding companies altered the industry's structure partly because they were able to weaken regulation's power.

Despite all of these flaws, state regulation was a success. Under it, the production and use of electricity grew and profoundly changed the US economy. The electric utility industry improved the productivity of manufacturing and enabled the creation of many new products and industries. For most living outside rural areas, electricity reduced household drudgery and improved standards of living. Despite these successes, the industry structure created by state regulation brought electric utilities continual controversy, proposals for reform, and a crisis that culminated in the 1930s with a law that again restructured the industry.

CHAPTER FOUR

Growth and Growing Pains

During the 1910s and the 1920s, electric utilities brought about enormous changes in the lives of most Americans and in the US economy. The use of electricity soared, and its price plummeted. Electricity became an indispensable part of the lives of nearly all those who did not live on farms. As its importance and size grew, the difficulty of the utility industry in developing efficient large-scale integrated networks became apparent. Two different public policy initiatives to address this problem failed, and that failure helped set the stage for the greatest political crisis the industry had ever faced.

Spectacular Growth

Total generation of electricity by electric utilities and number of customers grew rapidly before 1929. Electricity generation and number of customers in 1929 were 15.7 and 12.4 times their respective 1907 values, average annual growth rates of 12.5 percent and 11.4 percent (figure 4.1). In 1929, the average inflation-adjusted price charged residential users for electricity had fallen by more than 80 percent since 1907, while the average price for all uses fell by just under 47 percent. In both cases, nearly the entire price decline had happened by 1920. During the 1920s, there was little change in the average price for electricity (figure 4.2). Gross investment by privately owned electric utilities grew steadily until 1912. Six years later, gross investment had fallen almost to the level it had been in 1900. For the next six years, however, the industry experienced its most rapid increase in gross investment; by 1924 annual gross investment was more than six and half times that of 1918 (figure 4.3).

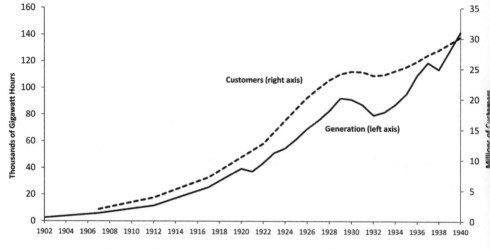

FIGURE 4.1. US electric utility generation and number of customers

Source: Cintrón and Edison Electric Institute 1995, 75

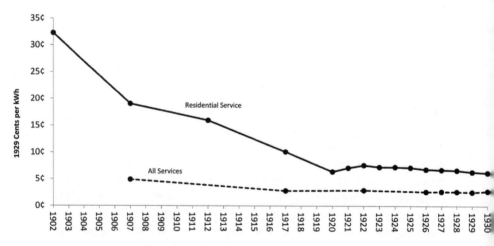

FIGURE 4.2. Price of electricity per kWh (1929 dollars)

Source: Carter et al. 2006, tables Db 234–37, adjusted by the CPI

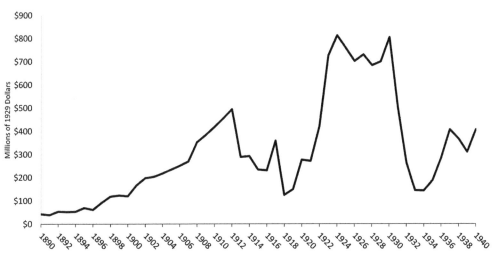

FIGURE 4.3. Real annual gross investment by US electric utilities

Source: Ulmer 1960, 320–21

Residential Use

In 1907, only 8 percent of all US dwellings had electrical service. By 1930, electricity was available in about 85 percent of nonfarm dwellings. The 25 percent of Americans living on farms in 1930 were not so fortunate. Only about 10 percent of farm dwellings enjoyed electrical service. For those with electricity, annual usage more than doubled to 547 kWh between 1912 and 1930. The average price of electricity to residential users in 1902 in current dollars was 16.2¢ per kWh, roughly equivalent to $4.44 in 2014 dollars. That price in the 1920s was between 6.3¢ and 7.5¢ in current prices, between 49¢ and 66¢ in 2014 dollars. By comparison, the average price for residential users in 2014 was 12.52¢.[1] Despite those price decreases, residential electricity users paid far more for their electricity than did businesses and other users—between four and five times the average price paid by industrial users.[2]

The development of household electrical appliances came quite early. The 1893 Chicago Columbian Exposition displayed electric fans, sewing machines, stoves, laundry machines, and irons. Such appliances were initially of interest only to the wealthy, who sometimes had their own generating plants. One factor that reduced the convenience of electric appliances was the lack of a system of plugs and outlets. Edison designed the

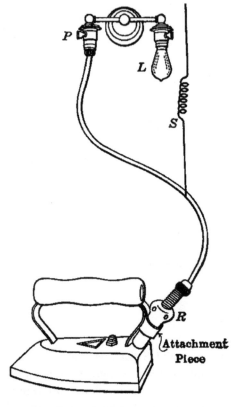

FIGURE 4.4. Connection of iron to lamp socket

Source: Keene 1918, 327

screw-in lamp-bulb socket that remains an international standard, but he did not design a wall socket for appliances. Access to electricity in a room was generally limited to light bulb sockets, usually placed high up and of-ten in the center of the room. Appliances had connectors designed to screw into a bulb socket after the bulb was removed (figure 4.4). Using an appliance could mean doing without lighting. Wall sockets designed for appliances as late as 1918 were of multiple designs, the most versatile of which simply replicated the light bulb socket (figure 4.5). The situation resulted in many different incompatible solutions; in 1915, the National Electric Light Association called for a standardized outlet, and two years later major manufacturers agreed on the system used (with modifications) in the United States today. The change did not occur overnight, however, and well into the 1920s, appliances were still available with incompatible plugs and light-socket plugs.[3]

As electricity prices declined, usage of appliances increased, despite the plug nuisance. Electric irons were initially the most popular of the electric gadgets. These replaced solid-metal irons (sometimes called sad irons) heated on a stove. Several such irons were required to replace those that cooled with use. The self-heating electric irons were quicker, easier, and avoided the large amount of heat generated by a stove. Data on the manufacture of electric appliances (with the exception of light bulbs) are generally unavailable prior to 1919. During the period between 1919 and 1929, the number of irons manufactured was usually more than twice that of any other appliance, over three million in 1929. As early as 1912, utilities

FIGURE 4.5. Appliance wall socket using a light bulb connector

Source: Keene 1918, 329

TABLE 4.1 **Annual manufacture of electric appliances**

Year	Heaters, air	Grills	Percolators	Toasters	Waffle irons	Disc stoves, hot plates, and ranges, under 2.5 kw	Domestic ranges, 2.5 kw and over	Fireless cookers, electric	Curling irons	Flatirons	Heating pads	Water heaters for permanent installation	Vacuum cleaners
1919	207,476									1,407,822			977,339
1921	310,363							2,899	227,220	1,146,179		36,672	739,534
1923	495,615	184,834	260,050	476,606	131,445	216,559		11,640	1,480,395	2,434,280		14,507	1,240,742
1925	317,851	209,318	434,095	735,856	315,777	255,587	85,158	63,201	1,534,786	2,936,361	265,588	54,042	1,107,592
1927		196,688	791,726	1,276,145	653,893		112,972	73,406	1,355,615	2,936,658	396,272	33,636	1,128,256
1929	503,494	262,564	709,154	1,612,790	839,633	694,230	225,477	79,945	1,334,242	3,132,882	529,733	58,302	1,382,070
Growth rates	11.1%	5.9%	16.7%	20.3%	30.9%	29.1%	24.3%	41.5%	22.1%	10.5%	18.8%	5.8%	3.5%

Sources: US Bureau of the Census 1924; US Bureau of the Census 1933, 1107/vol. 2, 1129

distributed irons on a free-trial basis with small monthly payments.[4] Large quantities of vacuum cleaners were made during this period, but the number manufactured each year showed relatively little growth. Appliances that grew substantially in popularity included curling irons and toasters (table 4.1).

A survey taken in Chicago in 1926 showed that 85.2 percent of the customers of the electric utility used electric irons, and almost 71 percent used vacuum cleaners. More than 30 percent of customers also used washing machines and toasters. Lighting, however, remained by far the dominant use for electricity in the home. A 1923 survey showed that 75 to 80 percent of residential electricity use in Chicago was due to lighting.[5]

Industrial and Commercial Use

The use of electrical energy in manufacturing increasingly replaced energy from steam engines. Electricity's share of the total energy used in manufacturing grew steadily from under 5 percent in 1899 to over 85 percent in 1929.[6] Since at least 1912, industrial use of electricity, including electricity from nonutility sources, exceeded all others (figure 4.6), and industrial use accounted for a large share of electricity sold by utilities. Of the electricity produced by utilities between 1926 and 1940, industry

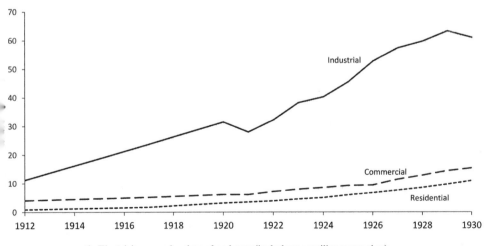

FIGURE 4.6. Electricity usage by class of end user (includes nonutility generation)

Source: Carter et al. 2006, tables Db 228–31

TABLE 4.2 **Total horsepower of electric motors used in manufacturing by source of electricity**

Year	Purchased energy	Self-generated energy	Self-generated as percent of total
1899	182,562	310,374	63%
1904	441,589	1,150,886	72%
1909	1,749,031	3,068,109	64%
1914	3,884,724	4,938,530	56%
1919	9,284,499	6,969,203	43%
1923	13,365,663	8,821,551	40%
1925	15,868,828	10,254,745	39%
1927	19,132,310	11,219,979	37%
1929	22,775,664	12,376,376	35%

Source: US Bureau of the Census 1933, vol. 1, 112

used between 46 percent and 57 percent of the total. Industrial use was dominant even earlier. In 1917, motors used almost two and a half times as much utility-supplied electricity as lights, and industrial users surely accounted for most motor use.[7]

Utilities were not the only source of electricity, and significant proportions of the electricity used by industry came from isolated plants. In the same 1926 to 1940 period, between 30 percent and 40 percent of all electricity used by industry came from isolated plants.[8] In earlier years, an even larger proportion of electricity was self-produced. In 1904, only 28 percent of electric motor power used in manufacturing came from utilities. This percentage increased steadily, reaching 65 percent by 1929 (table 4.2). The competition utilities faced from isolated plants at least partially explains why industrial users were charged a fraction of the price for electricity as were residential users, who had no choice but to get their electricity from utilities. Between 1926 and 1929, utilities sold over four times as much electricity to industry as to residential users, but since the ratio of the price charged residential users to industrial users was even higher, total revenue from residential users exceeded that from industrial users.[9]

The Impact of Electrification on Manufacturing

Electrification had a profound effect on the design and organization of factories.[10] Prior to electrification, the primary source of power for factories was steam power, with water power used to a lesser extent. Steam

provided a low-cost source for power, but distributing that power throughout a factory presented a problem. Large steam engines were more efficient than small ones, and each steam engine required a boiler and the fire
to heat the boiler, making it cheaper to use one large steam engine than
several smaller ones. A single steam engine (or water wheel) typically powered all of the machines used in a factory. The most common means for
distributing the power was "millwork," which mechanically connected each
machine in a plant to the single source of power through a system of ceiling
shafts, pulleys, belts, and gears. The steam engine powered a constantly
rotating ceiling shaft. A system of belts, usually leather, and pulleys connected each machine to either that shaft or another itself connected to the
main shaft by belts or gears. The machine operator would have a clutch
that allowed control over the connection of that machine to an overhead
shaft.

In order to withstand the strains associated with this power transmission,
the shafts had to be huge. In the 1890s, the transmission of 170 horsepower
across a distance of 200 feet required a solid steel shaft five inches in diameter, weighing over ten tons, supported by thirteen sets of bearings. Vibra-
tion in the shaft was unavoidable, and alignment problems recurred. The
need to support this monstrosity was a major factor in the design of manufacturing buildings.[11] A factory room could have dozens, even hundreds of
flapping belts that were a common cause of industrial accidents (figure 4.7).
The supporting bearings needed lubrication, often provided by drip oil-
ers, which required constant refilling and which slung oil throughout the
room. In a multistory factory, a belt, or system of belts, passed through
holes in the ceilings/floors separating stories, to enable the main shaft to
power similar ceiling shafts on all floors. The holes also enabled fire to
spread easily from floor to floor. The system of belts and shafts created a
manufacturing environment that was dark, dirty, and dangerous. Flexibility in machine placement was restricted because machines powered by the
shaft had to be located along it. Expanding or altering a factory was difficult, as was the use of multiple buildings at a single plant site. Energy losses
in such a system could be enormous, amounting to as much as 30 percent.
There were alternatives that would have avoided many of these problems,
including the use of distributed steam, hydraulics, and compressed air. The
last became widely used in mining, but millwork remained the dominant
system in manufacturing until electricity replaced it.[12]

Electrification did not immediately correct the problems with millwork. Initially, a large electric motor simply replaced the steam engine;

FIGURE 4.7. Factory using shafts and belts, Germany

Source: Sternegg 2014

the factory continued to use all of the same shafts and belts. Later, the use of several smaller motors (group drive) replaced that of a single steam engine. Each of these smaller motors would power a separate smaller shaft connected by belts to multiple machines in the group. Eventually unit drive was used, where each machine tool was powered by its own small electric motor.[13] Unit drive freed the overhead space required by shafts and belts for alternate uses, such as cranes to move material within the plant. The small motors brought greater flexibility to the placement of machines. The production process, rather than shaft access, could determine machine placement, and less total floor space was required. It became easy to locate different processes in different buildings and easier to modify the layout of a factory when needs changed. Expansion became simpler. Factories became cleaner and safer work environments.

Improving productivity means getting more output for the same input or inputs. One common measure of productivity is labor productivity, the amount of output per unit of labor. Since workers differ in their levels of skill and education, this measure requires adjusting the measure of labor used to account for these differences. Increasing other inputs to production, such as the amount of machinery (capital equipment) or energy used per worker, will enable more output per worker. Total factor productivity attempts to take all inputs to production into account. In principle, an improvement in total factor productivity indicates a cost-less increase in output. Its calculation requires numerous assumptions. To the problems of measuring labor input are added problems measuring all other heterogeneous inputs. Based on economic theory, a mathematical formula or set of formulas (the production function) models the relationship between the inputs to production and the outputs produced. Econometric methods enable an estimation of the values of the production function's parameters. When an estimated relationship from an earlier time is used to predict output for a later time, increased total factor productivity will cause actual output to be higher than predicted. The amount of extra output is a measure of the increased total factor productivity between the two periods. Something must have been responsible for the increased output that was not accounted for, but the analysis cannot identify what that was. It is reasonable and common to credit improved technology for the increase, but that must be conjectural.

The productivity estimates by John Hendrick published in 1961 remain the authoritative historical measures for the United States. He estimated labor productivity annually but estimated capital and total factor productivity for only selected years. As shown in figure 4.8, the rate of increase in total factor productivity had a sharp upward turn after 1919. Between 1869

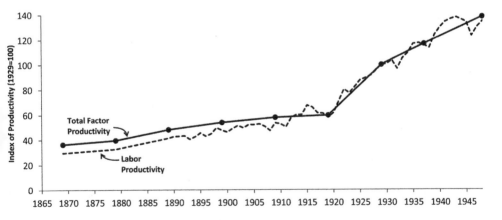

FIGURE 4.8. US manufacturing productivity

Source: Kendrick 1961, 464

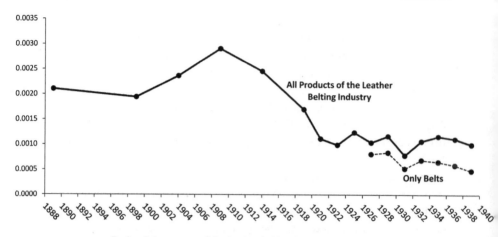

FIGURE 4.9. Ratio of the output of the leather belting industry to total value added in US manufacturing

Source: US Bureau of the Census, *Census of Manufacturers*, various years

and 1919, the annual compounded growth rate in total factor productivity was 1.0 percent. Between 1919 and 1929, that growth rate jumped to 5.2 percent. The annual data for labor productivity show greater fluctuation but closely track total factor productivity.[14]

In addition to improved factory organization, the 1920s saw other technological innovations in manufacturing, such as the assembly line, that likely improved total factor productivity. Electrification facilitated many of these innovations, and the argument that electrification was responsible for improved productivity seems compelling. The switch to electrification would have reduced the use of transmission belts in manufacturing. However, that drop occurred before the 1919 productivity growth increase. Figure 4.9 shows the ratio of the value of all products produced by the leather belting industry to total value added in US manufacturing and the value of just leather transmission belts after 1927.[15] The fall in the ratio between 1909 and 1923 is consistent with a shift to electric motors when plants were being constructed or altered and would lead a measure of any drop in the use of belts in all manufacturing. This rapid decline, however, stops after 1923. The lag between the shift away from belts in new facilities and the sudden rise in productivity is puzzling.

In addition to changing factories, electrification also changed manufacturing's labor force requirements. Between 1900 and 1940, the relative demand for unskilled labor and for workers with clerical and managerial skills increased, while the demand for workers with intermediate skills declined. This has been described as a "hollowing out" of the skills of the workforce.[16] Electrification enabled factories to reorganize and more easily employ techniques, such as assembly lines, that were probably most directly responsible for those changes in the workforce.

Interconnection and Large Integrated Networks

The magnitude of the benefits that were being lost by the industry's failure to use larger fully integrated networks became clear with World War I and led to two failed public policy proposals in the 1920s that would have restructured the industry. The failure of the first proposal clearly demonstrated the impediments to the creation of such networks that were imposed by the existing industry structure. The second proposed a more profound structural change that anticipated some of the modern changes in the industry.

During World War I, concern arose that problems with the availability of resources would hamper the war effort. This led to the creation of the War Industries Board. Several parts of the country experienced power shortages, which were investigated by the Power Section. It found cases where interconnection alone would have increased the availability of power without any increase in generating capacity. In the region around Pittsburgh, for example, interconnection would have reduced needed reserve capacity by 150,000 horsepower.[17] In one area of Ohio, interconnections would have permitted larger plants located near sources of condensing water to replace many smaller plants, saving costs and much needed coal then in short supply.[18] The government's role in evaluating the efficiency of the US electric utility industry brought increased awareness of the unrealized benefits of larger networks.[19] This led in the 1920s to two similarly named proposals, Superpower and Giant Power, that would have enabled larger networks by altering the industry's structure.

Superpower

The Superpower plan proposed a new integrated network serving all of Massachusetts, Rhode Island, Connecticut, and New Jersey, as well as portions of seven adjacent states (figure 4.10), an area then served by 315

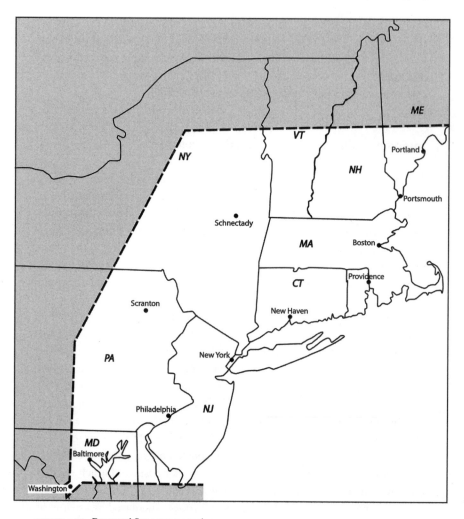

FIGURE 4.10. Proposed Superpower region

Source: Prepared by author based on Murray et al. 1921, map opposite p. 12

separate utilities, although the presence of holding companies makes this
an overestimate of the extent of separate ownership. The author of the
Superpower proposal, William S. Murray, was an electrical engineer and a
prominent advocate for the superiority of private over government ownership of electric utilities. His consulting work had included estimating
the benefits of utility consolidations.[20] He had promoted the idea of the
new network at professional engineering meetings and to executives of

utilities, railroads, and industries in the region. After endorsement by the secretary of the Interior, Congress appropriated funds for the US Geological Service to investigate the idea, and Murray was made the study's director. The study was completed in twelve months, and its final report was issued as a government publication in 1921.[21]

Starting with the characteristics of the region's electric supply system in 1919, the study's authors developed a transition plan for an optimal fully integrated network in 1930 based on already existing technology. In 1919, the region had 558 generating plants. Of those, 218 were to be retained, primarily for use during peak periods, and 55 huge new plants were to be built. The average capacity of the existing steam plants in the region was 10,000 kW. Those that were to be retained had an average capacity of 44,600 kW, while the average size of the new steam plants was to be 218,000 kW. Those new plants were to be located at the best sites in the region for coal and cooling water availability. Generation by hydroelectricity was to increase from 15 percent of the total in 1919 to 21 percent in 1930. The new plants were to be connected to population centers by a system of 8.8 thousand miles of new high-voltage transmission lines.

Even though conservative assumptions were generally used, the estimated benefits of the new network seem incredible. The fully integrated network lowered the total cost of capital equipment that would have been required in 1930 by 23.5 percent. Projected production costs per kWh within the new network fell by over 50 percent. The report added to these an estimation of the benefits the new network offered railroads and manufacturing from the perspective of those industries.

Extensive analysis led the study to conclude that 19,000 of the existing 36,000 miles of rail lines in the region could be economically converted from steam to electricity. A conservative assumption about the cost of wholesale electricity was used, but the railroads were assumed to bear all of the conversion costs, including the required new electrical infrastructure, the cost of new locomotives, and the costs arising from the early retirement of equipment including steam locomotives. The railroads' lower costs would have provided them with a 14.5 percent annual return on their investment.[22]

A special tabulation of the 1919 census of manufactures for the region provided data on the sources and uses of energy by fifty-three industrial classifications. These data were distributed to a number of individual engineers and manufacturers for their opinions about which energy uses could be profitably switched to electricity from the network. The proportion of the total energy use in 1919 that did not come from electric utilities was large. The analysis of the data in the study showed that almost 74 percent of manufacturing power came from sources other than utilities. Steam power accounted for part of this as did the 47 percent of the electricity used in electric motors. The study's engineers took into account the need for process heat by the different manufacturing sectors, and they assumed that all isolated plants used the best available technology. They concluded that about 70 percent of the capacity of prime movers (excluding water wheels) could have been economically switched to purchased electricity by 1930, and that the switch to burning coal in highly efficient utility plants would have resulted in a net savings of more than 25 percent of the region's total (not just in isolated plants) industrial use of coal.[23] The primary equipment cost to manufacturers would have been the quintupling of new electric motors. In many cases, there would have been a net benefit to some manufacturing firms of switching to utility-supplied electricity even without the new network. Those unrealized benefits should not have been counted as a benefit of the new network, and their inclusion may have exaggerated the estimate of the network's total benefit. That estimate was huge, and amounted to a 33 percent return on the total investment made by utilities, railroads, and manufacturers.[24]

The enormous benefits claimed for the new network were never disputed and were unrelated to its rejection. Implementation of the Superpower network required the creation of many new organizations that would have altered the structure of the region's electric utility industry. These organizations would have been necessary to finance and create the new facilities, plan for future network growth, and take control of day-to-day operation of the network. Some method would have had to be devised to manage and finance transitional issues, including the retirement of most of the region's generating facilities. The Superpower proposal did a magnificent job of addressing and answering the technical problems of implementing a fully integrated regional network, but it ignored the far more difficult organizational, institutional, and managerial problems. The only point at which the report touched on this was a statement by Murray that Superpower would never supplant or compete with the existing utilities; the existing structure would remain.[25]

In June 1921, Murray presented a preview of the report to the NELA including summaries of the appendices, which made up most of the report.[26] The actual appendices closely followed his previews except for Appendix A. In that appendix, Murray had promised to discuss the financial principles of the Superpower system and outline the necessary new legislation. The actual Appendix A, entitled "Organization," was an acknowledgement of various individuals and organizations that had contributed to the study. Before the report had been issued, the organizational problem was discussed by the Superpower Advisory Board, which consisted primarily of officials representing electric utilities and electricity-using industries but also included Secretary of Commerce Herbert Hoover.[27] The board urged the creation of a Superpower Corporation chartered by the federal government and subject, at least in part, to federal regulation, but it could not settle the issues of the authority the new corporation would have over the existing utility companies, the mechanism for operating the new integrated network, or the scope of federal regulation. A month before the report was completed, a meeting with utility executives sought to enlist their support. Instead, they raised serious objections involving corporate and financial, but not technical, issues.

After the report's submission, Secretary Hoover took leadership of the Superpower movement and invited the region's governors and members of regulatory agencies to a conference. A Northeast Super Power Commission was created, but it achieved little, despite Hoover's emphasis on voluntary agreement and his initial goal of simply increasing utility interconnection. Superpower remained a topic of discussion for decades, although the term sometimes referred to simple interconnection rather than full network integration. In 1925, Murray acknowledged that the business organization problem was more difficult than he had originally believed and that unified ownership of the utilities (by a holding company) was probably a necessary precursor to an integrated network.[28]

Superpower established that the replacement of many separate networks with a single integrated network would provide huge economic benefits to its region, and they were never doubted. Despite that, its creation by the voluntary action of the privately owned utilities in the area was impossible. Under the existing industry structure, voluntary consensus among the 315 utilities involved required an implementation plan that made it beneficial to each of the individual utilities. Superpower posed a significant risk to the individual utilities. Many current and future decisions had to be made that sought to benefit the entire region but that also involved transaction-specific capital equipment. An individual utility would have had little control over those decisions once an agreement had been made. Not only were the risks high, under regulation, utilities operated in an environment where their survival was virtually guaranteed by the prevention of competition and rates designed to ensure their continued viability.

Effective regulation would have ensured that much of the network's benefits would have gone to users, not to the utilities. The situation was one in which each utility felt the need to be cautious and to be skeptical of any proposed agreement. Murray was correct; the only way that many of the risks would have been eliminated would have been for all of the utilities to be under common ownership.

Those who believed government ownership of utilities was preferable to private ownership were skeptical of Superpower. Although it promised huge benefits, the only benefits measured were those that went to the utilities and to the business users of electricity. There were no attempts to measure directly the benefits to individuals either through lower rates or through increased access to electricity. However, even though the existing structure of privately owned utilities had prevented the realization of his dream, Murray remained a vocal and controversial advocate of the superiority of private ownership to government ownership.

Giant Power

Although the Superpower proposal shied away from matters of institutional structure, they were the heart of the Giant Power proposal. It proposed radically restructuring the industry to divide the ownership of the generation, transmission, and distribution among separate companies, anticipating an idea that resurfaced in the United States some seventy years later. Implicit in the Giant Power proposal was the notion that the existing structure of privately owned utilities had failed to provide the nation with many of the benefits promised by electrification. Under Giant Power, private ownership would have continued to dominate the industry, but government control would have substantially increased. The advocates of Giant Power also believed that the industry had not taken full advantage of existing technology. Superpower's technical proposals were based only on the use of technologies and practices already employed by domestic utilities. Giant Power proposed the use of technologies used in other countries but not in the United States. This opened it to criticism that those technologies were designed for the different conditions in foreign countries and would not work well here. Despite private ownership's continued domination of the industry under Giant Power, privately owned utilities in Pennsylvania and the rest of the country it saw it as a threat and vigorously opposed it. Murray erroneously described it as communistic.[29]

The key figures in the initiation of the Giant Power proposal were Gifford Pinchot, who became governor of Pennsylvania in 1922, and his utilities advisor, Morris Llewellyn Cooke. Both Cooke and Pinchot were members of the commission Hoover had formed to implement the Superpower proposal. Partly because Superpower gave no consideration to the unavailability of electricity in most rural areas, both came to view Hoover's commission and the federal government as championing the interests of privately owned utilities over the public interest. Progressive reform had to come from a state rather than the federal government. Giant Power took the form of a proposal from the governor's office to the Pennsylvania legislature.[30] Although that proposal covered only the state of Pennsylvania, the authors expected that a multistate network would eventually embody its principles, and an advisory board included members from other states.[31] Implementation of the Giant Power plan did not require achieving a consensus among existing utilities; its implementation depended on passage by the Pennsylvania legislature.

Pinchot was a Progressive Republican in the tradition of Teddy Roosevelt. A central concern of that group was the conservation of public resources. Without some type of government regulation, the profit incentive can be expected to encourage the wasteful overexploitation of those resources by private interests. The same concern about public resources played a major role in the development of hydroelectricity, discussed in chapter 7.

Pinchot spent time in Europe studying how public policy was used to manage forests consistent with long-run efficiency. He played a role in creating the Forest Service in the Department of Agriculture, and he persuaded his father to found America's first school of forestry (at Yale). Pinchot was not a socialist, but like some other prominent Progressive Republicans, his awareness of the issues involving publicly owned resources extended to a concern that the profit motive would also result in behavior by electric utilities contrary to the public interest. In his view, public policies needed to mitigate the power of private interests and subject them to continuous government oversight and regulation.[32] Pinchot recognized the enormous potential of electricity to better people's lives, but he felt the existing structure of privately owned utilities was falling short by denying electricity to rural areas and by overcharging residential users. Like others, he regarded the existing system of state regulation as ineffective at protecting the public's interest. The solution was not government ownership but a restructuring of the industry and improved regulation:

... a delay of even five years in establishing effective public control will bring
Pennsylvania and the Nation face to face with the immediate threat of an over-
whelming and almost uncontrollable electric monopoly. . . . The question before us is not whether there will be such a monopoly. This we cannot prevent.
The question is whether we shall regulate it or whether it shall regulate us. . . .
The unregulated domination of such a necessity of life [electricity] would give
to the holders of it a degree of personal, economic, and political power over
the average citizen which no free people could suffer and survive. . . . We and
our descendants [shall] be free men, masters of our own destinies and our own
souls, or we shall be the helpless servants of the most widespread, far-reaching,
and penetrating monopoly ever known. Either we must control electric power,
or its masters and owners will control us.[33]

In 1923, the Pennsylvania legislature authorized the creation of the Giant Power Survey Board, and Morris Llewellyn Cooke became the survey's
director. Cooke was a consulting engineer who shared Pinchot's Progressive concerns about privately owned utilities. As Philadelphia's director of
public works, he became involved in a very public battle with Philadelphia
Electric over the rates they charged both residential users and the city for
street lighting. Although he was unable to make the case that the value
Philadelphia Electric claimed for its capital equipment was excessive, the
utility ultimately backed down and reduced its rates to both groups. This
burnished his reputation among the critics of privately owned electric utilities, but among those utilities he became viewed as a dangerous radical. In
1935, President Roosevelt named him to lead the new Rural Electrification
Administration (chapter 8).

The objectives of the study given by the legislative act included surveying
the fuel and water resources within Pennsylvania available to generate electricity and developing policies to ensure an abundant and cheap supply of
electricity for use by industries, railroads, farms, and homes. The act also
specified a number of specific policies and technologies the study was to examine. Western Pennsylvania possessed one of the nation's richest sources
of bituminous coal, and the act directed the study to investigate the use of
large coal-burning generating plants located near the mines, the processing
and utilization of coal byproducts, and the development of multipurpose
water projects to control floods as well as generate electricity.[34]

In 1925, the legislature received the report of the Survey Board. Like
Superpower, Giant Power sought to create a large integrated network
spanning the service areas of existing utilities. Like Superpower, Giant
Power proposed replacing most existing generating plants with large base-

load plants that would supply users through new high-voltage transmission lines. In Giant Power, these new base-load plants would be located at mines in the state's coal mining region. However, that region lacked the cooling water required by power plants. The solution presented was to use cooling towers, which were used in Germany but not the United States. They were one of the controversial technologies attacked by the plan's opponents.

When the report was submitted, there were 187 local utilities operating in Pennsylvania, but nineteen different holding companies controlled the eighty-two of them that were responsible for over 96 percent of total generation. Murray saw holding companies as agents to bring about the type of integrated network conceived by the Superpower report. The Giant Power report discussed the difference between full integration and mere interconnection but criticized the role holding companies were playing in developing integrated networks. Holding companies only integrated those operating companies they owned, and they were often not contiguous. This sometimes led to a wasteful pattern where transmission lines would cross, but not connect with, the territories of operating companies owned by other holding companies.

A third of the report concerned the failure of electric utilities to serve rural areas, and it provided analysis showing that rural electrification provided positive benefits. Two new types of rural distributors were authorized: rural electric districts and rural electric cooperatives. Both would eventually become important, but their use in 1925 would have been innovative. In addition, municipally owned utilities would be free to extend their distribution systems beyond their boundaries to adjacent unserved areas. Even though regulation had prevented competition in the service area of an existing utility, the regulatory commission was given the authority to reassign any part of an existing utility's service area not provided with access to electricity. Techniques not then in use to reduce the cost of rural distribution were described. Examples of policies in other countries that had advanced rural electrification were presented, and the report advocated considering subsidizing rural distribution lines.

The Superpower plan failed because it took a minimalist approach to the problem of designing a new institutional structure. By contrast, Giant Power's approach was far from minimalist. It described in detail proposed changes in the industry's institutional structure. Existing utilities would become local distributors of electricity purchased from distant power plants owned by other companies and transmitted over the lines of a different set of companies operating as common carriers.[35] The existing Public Service

Commission would continue and would have new powers, including over the rates and financial securities of the new generation and transmission companies. In addition, a new regulatory agency, the Giant Power Board, would have extensive control over the planning and operations of the new companies and the new network.

Each Giant Power generating company would receive a fifty-year permit granted by the Giant Power Board. The board would have the final authority over the siting of each generating plant. Each company would have responsibility for the construction and operation of large (at least 300,000 kW) generating plants located in the coalfields of Western Pennsylvania. In addition to generation of electricity and its sale to distributing companies, the new companies were also to be in the business of mining coal and of producing and selling coal by-products. The companies would have the power of eminent domain to acquire coal-mining rights to meet their plants' needs for fifty years. At the conclusion of the fifty-year term of the permit, the board could renew it or, upon payment of the value of the company's investment, could assign it to another company or to the state government.

Giant Power transmission companies could apply for fifty-year permits to construct and operate transmission lines in locations determined by the board. These transmission companies would connect the Giant Power generating stations, and other generating plants, to the distributing companies and would operate as common carriers. Either the state or the transmission companies would own the land needed for the transmission lines, and the companies would be able to acquire the needed land using eminent domain. At the end of the fifty years, the board could renew or reassign the transmission permits on the same basis as the generation permits.

The proposal made changes to the system of regulation, including shifting the valuation of a utility's capital investment from "reproduction cost" to "prudent investment," which would likely have resulted in lower rates (see chapter 3). Regulators could explicitly take into account the quality of service a utility gave its customers in setting rates and grant higher rates to utilities with good service than to those with poor service even if their costs were the same.

Unlike Superpower, the battle over Giant Power occurred in the public political arena. Pinchot and Cooke were successful at getting favorable national publicity for the plan. Opposition on the part of existing utilities, however, was intense. The Pennsylvania legislature twice rejected Giant Power, and efforts to interest other states were unsuccessful. In the political climate of the time, the concerns of the existing utilities, and the widespread opposition to government interference in business, overcame concerns over the shortcomings of the existing industry structure. Although Giant Power was presented as a way to improve the industry's performance and efficiency, it was clearly predicated on the view that these improvements required an industry restructured to provide greater government control. Privately owned utilities were unable to agree on the institutional structure required for Superpower, but they easily came to a unified position on Giant Power—opposition.

Would the new institutional structure provided by Giant Power have been better than the one it would have replaced? Would it even have been workable? Could it have provided a path for the national industry? Although its advocates saw the principles of Giant Power as transferable to other states, many of its provisions were directly applicable only to Pennsylvania where major coal deposits were far from where electricity was needed. Details about some aspects of the new structure were not worked out, particularly if the lines of different Giant Power transmission companies were interconnected, which would have been unavoidable. As explained in the introduction, there are serious problems compensating multiple owners of lines in an interconnected transmission network. Another problem comes from the requirement that generation and use of electricity must match at each instant. Some authority must be responsible for dispatch, the continuous determination of how much electricity each generating plant should produce at each instant. The report did not provide details on how this would be done. Designing a new institutional structure for the electric power industry is difficult, and Giant Power's design was incomplete. Although ultimately unsuccessful, the Giant Power proposal is noteworthy in that it considered all of the functions needed to produce and distribute electricity. It was the first holistic approach taken by a proposed public policy to the redesign of the institutional structure of electric utilities.

Both Murray and Pinchot had foreseen another mechanism that could enable the industry to develop larger and more integrated networks: public utility holding companies. Their dominance brought both positive and negative changes to the industry's structure. It also created the greatest political crisis the industry had ever experienced.

CHAPTER FIVE

Public Utility Holding Companies

Opportunity and Crisis

Unlike state regulation, holding companies and the major changes they brought to the industry's structure were never the goal of public policy. However, the public policy that created state regulation was a major cause for the importance of holding companies. Holding companies were able to overcome the barriers state regulation created for the formation of larger fully integrated networks, and they brought other benefits to the industry. Holding companies were also able to manipulate state regulation in ways that undermined its ability to restrain utility profits, and they were able to conceal from investors the true risk of their investments. Their growing power and influence increased the old controversy over the role of profit-seeking companies in the industry and led to the most exhaustive federal investigation of any US industry. Events during the Great Depression humbled the once-powerful holding companies and resulted in public policy that led to their destruction and a restructuring of the industry.

Characteristics of Utility Holding Companies

The only assets of a pure holding company are the financial securities of other corporations. Unlike an investment company, such as a mutual fund, a holding company owns a large enough share of the voting stock of another corporation that it can control, or at least influence, the operations of that corporation. A pure public utility holding company produced no electricity and did not directly serve any customers, but it controlled operating utilities that played that role. There is no clear threshold of the proportion of a company's voting stock required for control or influence. Ownership of a majority of such stock is clearly sufficient, but a much smaller proportion may also be sufficient, depending on the number of other stockholders and the size of their individual holdings. A company with minority ownership sufficient to exert control may not have done so, complicating determining whether it was a holding company.

Prior to 1889, the legal authority of one corporation to own stock in another was limited. New Jersey then began issuing charters permitting such ownership and became a favorite state for corporate registrations. The fees that registration gave the state caused others to rapidly change their laws. In 1892, the first utility holding company was established.

The holding company provided a uniquely convenient form for consolidating local utilities.[1] This could be done without publicity, even if the utility's existing management was uncooperative, through the quiet purchase of stock. This consolidation enabled holding companies to join the service areas of the once-separate utilities into larger integrated networks; however, holding companies typically controlled far-flung operating companies too distant to be integrated. Holding company control grew during the 1920s until they controlled over 80 percent of total generation, and that control became increasingly concentrated. The bottom-level holding companies in a system would be the ones owning stock in operating companies, but they would be controlled by higher-level holding companies often themselves controlled by still higher-level companies. This pattern obscured the size of any one system. The lack of transparency over actual control contributed to public concern.

Some holding companies did not have many of the troubling characteristics that characterized others. For them, the holding company structure was used only as a convenient way of forming large fully integrated networks. Their acquisitions were limited to operating utilities with adjacent service areas. Their systems were the equivalent of a single operating company. They had simple organizational structures and usually conservative finances. Legal convenience caused them to retain the status of holding companies, but they had little difficulty reorganizing as single operating companies. James B. Duke founded a number of utilities in Piedmont North and South Carolina that were owned by him, his family, and his associates. In 1917, Duke Power was formed as a holding company with control over all of those companies. In 1927, most of the once-separate companies were merged into a single operating company, but some subsidiaries remained separate. More than half of Duke Power's stock was owned by two stockholders, effectively protecting it from being taken over by another holding company.[2] Consolidated Gas Company of New York was both an operating and a holding company that acquired most of the electric and gas utilities in New York City and some adjacent areas. Many maintained a legally separate identity, but their operations were fully integrated.[3] In 1936, Consolidated Gas changed its name to Consolidated Edison. Some holding companies that created similar large integrated networks were later acquired by other holding companies with other, noncontiguous service areas.

Origins of Public Utility Holding Companies

To obtain the money required to acquire the necessary land and equipment, a utility would sell financial securities to outside investors. These securities fell into three broad classes—bonds, preferred stock, and common stock—differing in risk, expected returns, and the right to vote in elections for the board of directors. Holders of securities with voting rights had ultimate control over the company's policies and operations. Significant differences existed in the detailed characteristics of the securities within each class.[4] All securities could be sold to others, but the prices received might be different from those originally paid.

Bonds promised investors periodic fixed payments and return of their original investment after a specified number of years. If the company's income was insufficient to pay all investors, bondholders had priority. Bondholders did not have voting rights, but if the company defaulted on their payments, they had considerable recourse. Specific assets might be seized. The company might be forced into bankruptcy or receivership, usually resulting in the dismissal of top management. Receivers appointed by courts were put in charge of the company's operations with the goal of recovering as much of the amounts owed bondholders as possible. Bonds were the safest class of security. This made them easier to sell, and their safety enabled them to pay lower returns than other financial securities. This made them attractive to the company, but excessive reliance on bonds made them riskier because even a small fall in earnings might cause default. After the 1893 panic, many investors adopted the rule of thumb that no more than half the money raised from investors could come from bonds. Other securities would provide a cushion protecting payments to bondholders.

Owners of preferred stock were promised a periodic fixed payment, similar to an interest payment but called a dividend. Some preferred stock gave their holders some recourse if payments were not made, but they could not force the company into bankruptcy. Preferred stocks had no fixed term, but the company usually had the right to repurchase them at a fixed price. Preferred stockholders rarely had voting rights. Preferred stock was a riskier investment than bonds, but its expected return was higher. Preferred stock dividend payments had second priority to bond payments. Overreliance on bonds and preferred stock increased the risk of preferred stock. Common stock provided its cushion.

Common stockholders were not promised any return, but they had a claim to all of the company's profits after payments to bondholders and preferred stockholders. Any fall in profits would come entirely out of their returns, but there was no upper bound to what they might receive. They usually had all of the voting rights, although those were sometimes limited by the existence of different classes of common stock. The prices of common stocks were more volatile than those of other securities, but increases in their prices could give stockholders huge capital gains. Voting rights made common stocks the primary assets of holding companies. Common stocks were the riskiest financial instrument.

Holding companies improved utilities' access to financial markets and often provided technical and management support. This was particularly important for the earliest holding companies. Over time, other causes became responsible for the creation of holding companies. The stories of four different holding companies illustrate the different causes.

Electric Bond & Share[5]

Until 1915, utilities had to spend more on equipment than they received in revenue. The growth of a small local utility could be constrained by the need and difficulty of selling financial securities, especially common stock. Equipment manufacturers sometimes had to accept a utility's common stock as partial payment in order to make a sale. Those manufacturers faced the same problem as the utilities in converting those stocks into the cash they needed. In 1905, General Electric (GE) developed a method to solve that problem and create a more sustainable model for supporting the industry's financial needs. GE created a wholly owned, but independent, subsidiary, Electric Bond & Share, under the control of S. Z. Mitchell, and transferred to it the bulk of its utility securities. In December 1924, on the eve of a federal investigation into control of the power industry, Electric Bond & Share became independent when General Electric distributed its subsidiary's stock to its own stockholders.[6]

Mitchell's goal for the new company was to develop a system of financing that met the condition that no more than 50 percent of any company's capitalization came from bonds. This was done through the creation of several subholding companies. Electric Bond & Share would give them the operating company securities it held in return for a controlling share (but less than half) of their common stock. These subholding companies acquired additional operating company stock, including that of other utilities, until they held a controlling share (often 100 percent) of the voting stock of each company. In some cases, that control went through another holding company, a sub-subholding company. The holding companies' purchase of common stock enabled the operating companies to keep bonds' share of total capitalization below 50 percent. Because of their association with Electric Bond & Share (and General Electric), investors were willing to purchase the common stock and the bonds of the subholding companies. The system eased the struggle the local operating utilities had had with financing. GE benefited from the preferential treatment these utilities gave to its equipment.[7]

It soon became apparent that investor faith in a system's securities was greater if a holding company provided expert business and technical support to operating companies. American Gas & Electric, the first subholding company, created a management staff to provide these services to its operating companies. Electric Bond & Share then created a similar staff to provide those services to the operating companies of all its other subholding companies. For those operating companies, Electric Bond & Share maintained continuous oversight, performed all of the needed construction, and provided centralized purchasing. Operating companies were charged fees for these services, and these fees became a major source of profits for Electric Bond & Share. Many other holding company systems adopted this model of selling services to their operating companies.

Electric Bond & Share brought about consolidation of some local companies, but the operating companies distributed to each subholding company were located in several different regions of the country. Complete consolidation of any of the subholding company systems was not possible. In some cases, the operating companies of one subholding company were adjacent to those of a different subholding company, an arrangement that inhibited consolidation.[8] This pattern of each holding company controlling utilities was purported to give investors geographic diversity protecting them from adverse local conditions. Other holding company systems whose operating utilities were separated made the same claim, but the diversification effects were slight.[9] The nature and growth in the territories served by Electric Bond and Share are shown in figures 5.6 and 5.7.[10]

Stone & Webster[11]

Electric Bond & Share arose in response to financial, technical, and management needs of operating companies. Holding companies were not the only method of meeting these needs, however. Stone & Webster was an example of a company long involved in providing such services to the utility industry without being a holding company. The firm began in 1890 as a partnership of consulting engineers. Its original business included the design of generating and other facilities used by utilities. This was extended to supervising construction crews, a task made more effective by using its own employees and providing complete construction services to its clients. The young firm was hired by a group of bankers to advise them as to why the financial performance of a utility whose securities they held was so poor. Stone & Webster's report resulted in the bankers asking the company to take over the management of the utility. The 1893 panic brought financial distress to many operating utilities. This provided a speculative opportunity to purchase a utility at a depressed price, improve its operation, and sell it at a profit. Bankers, including J. P. Morgan, hired the firm to appraise the value of these distressed utilities and even assume their management. Stone & Webster came to manage many operating utilities under contract and had departments for auditing, preparation of reports, company record keeping, and marketing.

In managing utilities, Stone & Webster naturally became involved in the problem of financing. Initially it simply took the securities of the firms it managed and endeavored to sell them at the best price. Eventually it would combine with banks to underwrite securities, guaranteeing the utilities a minimum price. Later it organized a secondary market with a number of banks that would purchase securities when necessary to maintain a minimum price and resell them later at a profit. Stone & Webster thus came to function as an investment banker. Insurance was ultimately added to the portfolio of services offered. Stone & Webster came to offer all of the services provided by a holding company without actually becoming one. Sometimes it made relatively small investments in some of the properties it managed, but, except for firms purchased for resale, these were not large enough to give it the legal control possessed by a holding company. For those operating companies, Stone & Webster would select and train the local manager and assistant treasurer, and a Stone & Webster district manager and division manager would supervise both. Although on the payroll of the local company, Stone & Webster determined the continued tenure of the two local officials. They were effectively employees of Stone & Webster.

Although the most prominent, Stone & Webster was not the only company prior to the middle of the 1920s that offered similar consulting services, and they came to control substantial amounts of the nation's generation through management contracts. Their existence shows that holding companies were not necessary to serve the financial, technical, and managerial needs of many local utilities. For those concerned about the concentration of control of the nation's electricity generation, these consulting firms were seen as no different from holding companies. By being able to extend common ownership over adjacent utilities, holding companies had an advantage over consulting firms in the formation of large integrated networks. Consulting firms may have also played a role here, however. One government report stated that financing services, such as those provided by Stone & Webster, "undoubtedly hastened consolidation."[12] Improved access to financing enabled one firm to purchase another, but holding companies were better suited to this task, particularly when combining the operations of adjacent operating companies in different states.

As holding companies gained control over an ever-increasing portion of the nation's local utilities, Stone & Webster's business model became untenable. When a holding company acquired an operating company that had been a client of Stone & Webster, its business was lost to the holding company's servicing arm. To protect its relationships with operating companies, Stone & Webster was forced to create its own holding company, although it continued offering services to other operating companies. Increasing control of the industry by holding companies led Stone & Webster, in 1929, to merge with its holding company. Holding companies made it untenable for independent consulting firms to provide service to operating companies. Other consulting firms also created holding companies.

The United Corporation[13]

Bankers had long played a primary role in the formation of some public utility holding companies, but most were created by individuals with strong engineering and utility backgrounds. When a holding company acquired other companies, or when a holding company was created, new securities were issued. The strong investor interest in these securities, and the fees that came from placing these new securities, made new public utility holding companies attractive to investment bankers. By the time bankers assumed the major role in the formation of holding companies, however, few operating companies remained that would benefit from being part of a holding company system. Almost every operating company outside of a major metropolitan area was already part of a holding company system. With only a few available operating companies left, the new holding companies took to acquiring other holding companies. The resulting competition led to excessive prices being paid for these companies.[14] There still were opportunities to increase profits by creating larger integrated networks or by increasing monopoly power, but competition among holding companies made the former difficult because it was hard to acquire control of operating utilities with contiguous service areas. Nevertheless, investment bankers continued to sponsor new holding companies, and "superholding" companies emerged that acquired control over holding companies that had been at the top of their own large systems.[15] The most prominent example of this was the United Corporation.

The United Corporation, formed in 1929, acquired significant minority ownership in five top-level holding companies whose operating companies served a huge portion of the eastern United States, including large parts of Alabama, Georgia, Michigan, Mississippi, Ohio, New York, and Tennessee. Some of the strongest banking companies, primarily J. P. Morgan and Company, Bonbright and Company, and other affiliated banking interests jointly created the new company. It was the largest venture ever by J. P. Morgan (father or son), with a capitalization twenty times that of United States Steel.[16]

Five of the eleven directors of the United Corporation came from United's main subsidiaries and five came from the Morgan and Bonbright banks. The complex web of stock ownership among companies in the system (figure 5.1) suggested a "community of interest," creating a degree of joint control by bankers that exceeded that conferred by the direct ownership

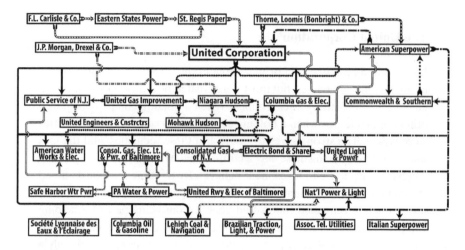

FIGURE 5.1. Stock ownership relationships among investment bankers and utility holding companies affiliated with the United Corporation

Source: See note 17

of stock by one company in the system by that of another.[17] Within a few months of the creation of the United Corporation, many of the same banking interests formed the Commonwealth & Southern Corporation by acquiring three other holding companies, sometimes to the surprise of the executives of the acquired companies. Although the United Corporation directly owned less than 4 percent of the stock of Commonwealth & Southern, other banking interests also held its stock, and the United Corporation effectively controlled the new holding company. The United Corporation also held small amounts of stock in other utilities that it did not control, but that caused speculation that its plans included eventual control of those companies.[18]

The holding companies formed earlier generally exercised direct control over the operating companies in its system. Superholding companies generally left most operational decisions to their subholding companies but did exert influence over their policies. The United Corporation, for example, forced two of its subsidiary holding companies to reform their system fees.[19] There was speculation that the United Corporation would eventually form a "Superpower" company by integrating the operating companies in its system, but during its limited existence, no movement in this direction occurred. This potential network (or networks) would have been huge, encompassing a large territory and providing over 20 percent of all the electricity produced by privately owned utilities. The territory served by the United Corporation in 1932 is shown in figure 5.7.

The Insull Group[20]

Although not the largest, no other holding company was to receive the amount of public attention as that of the "Insull interests," or the "Insull group." Samuel Insull, already mentioned in previous chapters, was probably the best known utility executive in the country and was certainly one of the most interesting entrepreneurs in the history of American business. Insull was willing to innovate in ways that involved large business and technical risks, and his methods came to be adopted by others.[21] He brought substantial benefits to the entire industry and to users of electricity. He did not shy away from publicity and became the industry's leading spokesperson. Most entrepreneurs achieve little or no success, a few achieve spectacular success, but Samuel Insull managed to achieve both spectacular success and spectacular failure. At its height, his personal fortune exceeded $150 million ($2 billion in 2014 dollars). Within a few years, he was "too broke to be bankrupt."[22] By the end of his life, he became a symbol of corporate evil, the most reviled individual in his industry.[23]

Insull came from England as a teenager to become Edison's personal secretary. He quickly became indispensable to Edison and was put in charge of Edison's manufacturing operations. After AC triumphed over DC, and his manufacturing business was taken over, Edison ceased to be involved with the electric industry he had done so much to create. Although Insull was made a vice president in the newly formed General Electric, with Edison gone he did not intend to remain in that position. The thirty-two-year-old Insull in 1892 accepted a two-thirds cut in pay to become president of Chicago Edison, one of about thirty small electric utilities serving that city. As a result of his promotional, business, and political skills (and luck), by 1907 he had consolidated all of Chicago's utilities into a single company, Commonwealth Edison, the world's first giant electric utility serving a large metropolis, and then duplicated this feat with gas utilities and intraurban transit. He brought tremendous operating efficiencies and lower prices. He made Chicago the most electrified city in the world. Insull also gained control of a large number of small utilities serving adjacent communities and, in 1911, formed a holding company, the Public Service Corporation of Illinois, to control and consolidate the operations of these companies.

Insull's reputation as utility wizard led the stockholders and creditors of poorly performing utilities in other parts of the country to seek his help, reminiscent of the experiences of Stone & Webster. As incentive or compensation, he frequently received securities in these companies. Initially he would put these companies on sound footing and then sell the securities, but he later started retaining them. In 1912, he formed a new holding company, Middle West Utilities, and held a controlling share of its common stock. The new company issued its own financial instruments, paid cash to Insull for his operating company securities, and provided those companies with the same range of services as other early holding companies.

Middle West Utilities grew rapidly during the 1920s, acquiring operating companies and other holding companies in an often-uncoordinated way, sometimes at what were likely excessive prices. In order to get desirable properties, Middle West was willing to accept undesirable ones. It came to control 239 operating utilities, twenty-four holding companies, and thirteen nonutility companies in 5,300 communities in thirty-six states.[24] This rapid growth contributed to a complex pyramided structure. At least one layer of subholding company lay between Middle West and all of its operating companies, but sometimes there were as many as four layers separating the top holding company from the operating companies—six layers in all (figure 5.2).[25] The system was so complex that GE's president defended the honesty of Samuel Insull with the assertion that he did not and could not understand his own system.[26] The nature and growth in the territories served by the Insull group are shown in figures 5.6 and 5.7.

Until 1929, no single holding company controlled all of the companies in the Insull group, which included the legally separate Commonwealth Edison, Middle West Utilities, and Public Service Company of Illinois, but the same people had controlling interests in these companies and were directors of all. Concerned that his companies had become a target for hostile takeover, in 1929 Insull created two new holding companies to hold controlling ownership interests in all of the companies in the group. These new holding companies incurred huge debts to finance the required stock purchases. The continued deterioration of the economy after 1929 caused earnings to fall, and the debt payments could not be made. Forty-one companies in the Insull empire became bankrupt. The nation had never before experienced a business failure of this magnitude.

Insull may have been the first to use the "customer ownership" plan (described in chapter 6), in which system securities were sold to the residential customers of its operating companies. When the Insull group ran into

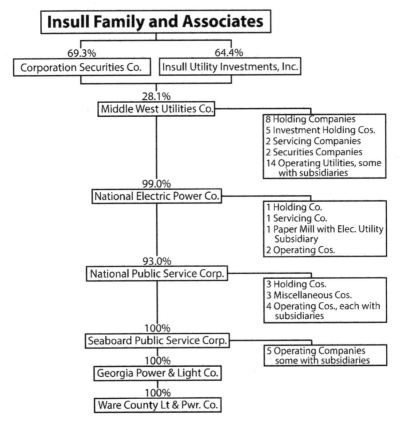

FIGURE 5.2. Holding and subholding companies for one operating company in the Insull system with percentage ownership of voting stock

Source: See note 25

financial difficulties, it attempted to shore itself up by increasing the sales of securities in small lots to its customers. The economic pain of the system's collapse was felt by an especially large number of people; more than a million held Insull securities.[27] The total losses by public investors was over $638 million ($11 billion in 2014 dollars), and many of these were middle- or lower-middle-class Americans.[28]

The fury against Insull following the collapse of the system was intense. "Insullism" entered the language as a synonym for the entire set of abuses of which holding companies were accused.[29] In his first campaign for the presidency, Franklin Roosevelt was milder than many when he criticized "the lone wolf, the unethical competitor, the reckless promoter, the Ishmael or Insull, whose hand is against every man's . . ."[30] Insull fled to Europe before he was indicted on several criminal charges. At a time when Europe was careening towards war, Insull effectively resisted intense extradition efforts, which were closely reported by the American press. His ultimate capture, return, and subsequent trials were major news stories. In a widely publicized federal trial, the jury quickly acquitted him of mail fraud, and all other charges against him were either dropped or ended in acquittal. This did not exonerate his reputation, however; he remained a symbol of corporate greed. No longer followed by the press, he died in 1938 of a heart attack in a Paris subway with twenty cents in his pocket.[31]

The Attraction of Holding Companies to Investors

The early holding companies were generally quite profitable. Between 1910 and 1925, the earnings of two of Electric Bond & Share's subholding companies grew over eighteenfold, similar to that of other holding companies.[32] Between 1925 and 1929, twenty-five new top holding companies, and hundreds of subholding companies at various levels, were formed.[33] Investors were bullish on utility stocks, particularly holding company stocks, during the 1920s. In 1939, Alfred Cowles calculated sixty-nine industry common stock indexes for the period 1871 to 1937, including one for utility holding companies and another for utility operating companies. Most of the industry indexes started later than 1871, some as late as 1933. The twenty industry indexes with the highest returns for the period from January 1920 to September 1929 (the month before the stock market crash) are shown in table 5.1.[34] Utility holding companies top the list with an annualized return of over 36 percent, corrected for deflation during the period. The end of the 1920s saw "The Great Bull Market," when stock values experienced spectacular increases. Table 5.2 shows the annualized returns from January 1928 to September 1929 for the top twenty industry indexes then available. Utility holding companies are third on that list and provided an annual return of over 75 percent.

Those two periods were not the only ones that provided investors in holding company stocks with impressive gains. Figure 5.3 shows the annual return to that index and three others for every twelve-month period from January 1920 through December 1929 plotted against the period's end.[35] For example, the chart shows a value of about 31 percent for the holding

TABLE 5.1 **Cowles industry stock indexes with the twenty largest annualized returns, January 1920—September 1929**

Rank	Index	Annual rate of return
1	*Utilities—electric, gas, etc.—holding companies*	*36.5%*
2	Retail trade—5¢ to $1 chains	34.5%
3	Electrical equipment	28.8%
4	*Utilities—electric, gas, etc.—operating companies*	*27.7%*
5	Automobiles and trucks (General Motors)	26.8%
6	Retail trade—department stores	26.8%
7	Office and business equipment	26.5%
8	Chemicals	26.4%
9	Miscellaneous manufacturing	25.4%
10	Agricultural machinery	25.4%
11	Retail trade—chain stores	25.3%
12	Lead and zinc	23.3%
13	Food products—other than meats	23.2%
14	Mining and smelting—miscellaneous	22.4%
15	Building equipment and supplies	21.3%
16	Utilities—telephone and telegraph	21.1%
17	Retail trade—mail-order houses	20.8%
18	Steel and iron (US Steel)	19.2%
19	Copper and brass	18.9%
20	Automobile parts and accessories	18.5%

Source: Cowles 1939, 167–268, and see note 34

TABLE 5.2 **Cowles industry stock indexes with the twenty largest annualized returns, January 1928—September 1929**

Rank	Index	Annual rate of return
1	Radio, phonograph, and musical instruments	85.8%
2	Airplane	75.8%
3	*Utilities—electric, gas, etc.—holding companies*	*75.2%*
4	Electrical equipment	62.4%
5	Office and business equipment	61.2%
6	*Utilities—electric, gas, etc.—operating companies*	*56.4%*
7	Retail trade—mail-order houses	54.1%
8	Copper and brass	51.1%
9	Mining and smelting—miscellaneous	50.8%
10	Chemicals	48.8%
11	Containers (metal and glass)	46.4%
12	Agricultural machinery	45.1%
13	Automobile parts and accessories	41.8%
14	Utilities—telephone and telegraph	39.6%
15	Miscellaneous manufacturing	39.2%
16	Retail trade—department stores	37.2%
17	Steel and iron (US Steel)	34.7%
18	Steel and iron (excluding US Steel)	34.3%
19	Miscellaneous services	33.6%
20	Oil producing and refining	32.0%

Source: Cowles 1939, 167–268, and see note 34.

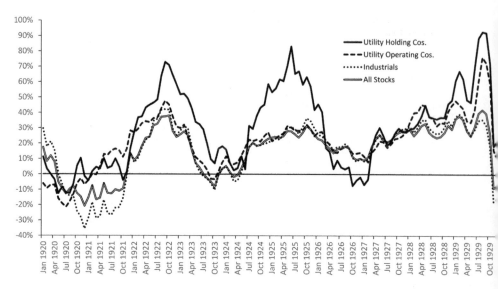

FIGURE 5.3. Return to selected Cowles indexes held for one year by end of period

Source: Cowles 1939, 168–71, 188–89, 198–99

company index for July 1924. This was the return to a portfolio of holding company stocks purchased in July 1923 and sold in July 1924. Holding
companies outperformed the indexes for industrial stocks and for all stocks
in ninety-four of the 120 monthly periods. The average annual returns to
the holding company index exceeded that of the industrial index by over
15 percentage points.

An imperfect measure of whether a stock may be overvalued is the ratio
of the price of the stock to the stock's share of the company's current earnings. If investors expect earnings to be stable, the price of stock for a company with a high price/earnings ratio would be overvalued. Suppose an investor with these expectations had a choice between one stock that had a
price/earnings ratio of 20 and another that had one of 100. The earnings of
the first stock are 5 percent of its value while those of the second are only
1 percent. The first would be more attractive; the price of the second would
be too high. If, however, the investor expected the earnings of the second
company to grow more rapidly than that of the first, its return would eventually exceed that of the first, and its stock would not seem overvalued. If
earnings did increase, and investors expected that to continue, so would
the price of its stock. An investor who purchased the stock earlier and later
sold it would receive a capital gain.

A high price/earnings ratio could also occur if there were a speculative bubble, and that could potentially affect many, or all, stocks. This could happen if investors in the past had enjoyed large gains in the value of stock. If the stock price increased over a long enough period, investors might conclude the price would continue to rise without a realistic expectation of continued earnings increases. That speculative behavior would cause the stock's price to continue to rise, fulfilling that expectation, and leading to an upward spiral in the stock's price. Eventually, the increasing gap between the stock's price and reasonable expectations about future earnings would lead to the realization that its price was too high. Some investors would sell the stock, and its price would fall. Other investors fearing that fall would continue would panic and sell their stock. The price would quickly collapse, with that fall occurring much more rapidly than its previous rise. A bubble pops faster than it fills. As early as 1925, some prominent figures in the electric utility industry were warning that speculation had led to excessive prices of utility stocks. Given the experience of the next five years, that view seems either premature or prescient.[36]

For much of the time when the price of a stock is rising rapidly, it is difficult to determine whether that rise was justified by expected future earnings or is the result of a bubble. Stock prices rose during the 1920s until September 1929, when they began a rapid fall. That pattern is suggestive of a bubble, but the actual cause remains controversial among modern economic historians.[37] One of America's greatest economists, Irving Fisher, famously argued days before the stock market crash that the large increases in stock prices were justified and not caused by a bubble and that price/earnings ratios were justified.[38] Among the Cowles indexes, holding companies had the eighth highest price/earnings ratio (31.0) for the entire year of 1929, and utility operating companies were twelfth (25.6).[39] At the end of August 1929, eight of the ten companies in the S&P composite index with the highest price/earnings ratios were utilities. The single company with the highest price/earnings ratio (105.4) was American Power & Light, an Electric Bond & Share subholding company.[40] Perhaps the expectation that holding companies would eventually create large fully integrated networks justified the high values investors placed on utility stocks.

Federal Investigations of Holding Companies

The lack of transparency in holding company systems created unease sufficient to cause several federal investigations. In the 1920s, some critics of the holding company system maintained (incorrectly) that a sinister "power trust" had already gained monopoly control over the nation's electric utilities. As the influence of these critics increased, so did the scope and intensity of the investigations. These culminated with an exhaustive investigation between 1929 and 1935 by the Federal Trade Commission. No other US industry has been subjected to the same degree of scrutiny as public utility holding companies.

Because hydroelectric facilities were often located on federal land or affected interstate navigation, hydroelectricity was a special concern of the federal government. The first investigation of the concentration of control among electric utilities reported that in 1911 General Electric interests controlled about 28 percent of the nation's total hydroelectric generating capacity.[41]

Continued concern about industry control resulted in a 1915 Senate resolution directing the Department of Agriculture to again investigate hydropower and report "any facts bearing upon the question as to the existence of a monopoly in the ownership and control of hydroelectric power in the United States."[42] Although the resolution referred only to hydropower, the report looked also at steam power. The authors attempted to deal with the inadequacies of the available data by using two different analysis techniques. The first determined which companies had direct control of an operating utility, defined as owning a majority of its stock, operating the utility's facilities through a lease, or managing it by contract. The stock ownership requirement was too stringent; a minority stockholder could be in control of a utility. Furthermore, only holding companies that directly owned operating company stocks were examined. Those holding companies, however, might have been subholding companies in large systems. Electric Bond & Share, for example, was not counted as controlling an operating company. This measure showed ten companies controlling about 23 percent of the nation's generating capacity. That certainly underestimated the industry's concentration of control.

The second analysis assigned operating companies to groups that included all those in the first analysis and also higher-level holding companies and investment bankers. The names of directors and principal officers of public corporations were easily obtained and a company was placed in a group if it shared a director or principal officer with at least one other company in the group. Sixteen groups were formed, and the total generation of the operating companies in each group was determined. By this measure, the group with the greatest control was the one containing General

Electric, accounting for about 23 percent of the nation's total generation. This approach had serious flaws. Not all of the companies in the same group would have been under the same control. Furthermore, individual companies were assigned to multiple groups. This caused the sum of the generation of all groups to be over twice that of the entire country.

Holding companies prospered and grew during the ten years from 1915 to 1925. Gross earnings of most of the holding companies that existed in 1915 grew at an annual rate of over 15 percent, reflecting both growth in the original operating companies and the acquisition of new companies.[43] This success encouraged the creation of new holding companies. At least thirty-four new holding companies received charters between 1923 and 1929.[44] Political concern over the control of the nation's electric supply also increased from those suspecting that control of the industry was becoming concentrated in holding companies whose power and wealth threatened private enterprise and exploited consumers. These companies were also seen as using a sophisticated propaganda operation to fool the public. George W. Norris, a Progressive Republican from Nebraska, became the leading spokesperson for those concerns. He contended that holding company power overwhelmed that of state regulation: "[state commissions could] no more contest with this gigantic octopus than a fly could interfere with the onward march of an elephant."[45] In 1925, he asserted in the Senate "practically everything in the electrical world is controlled either directly or indirectly by a Gigantic Power Trust."[46] Concern over monopolies had already resulted in a resolution asking for an investigation of the American Tobacco Company. In 1925, by a narrow vote Congress amended that resolution so that it also directed the Federal Trade Commission (FTC) to undertake a study:

> [to determine the] extent the General Electric Company, or the stockholders or other security holders thereof, either directly or through subsidiary companies, stock ownership, or through other means and instrumentalities, monopolize or control the production, generation, or transmission of electric energy or power. . . . and to report to the Senate the manner in which said General Electric Company has acquired and maintained such monopoly or exercises such control in restraint of trade or commerce and in violation of law.[47]

The resolution passed despite the fact that a little over a month earlier, doubtless to avoid any negative effects of this study, General Electric divested Electric Bond & Share, thereby relinquishing control over most of its operating companies. The FTC noted the divestiture, but prepared the report as of December 1924, before the divestiture. Although the resolution mentioned only General Electric, determining the extent of GE's control required an investigation of the entire industry. The study also investigated the concentration of control in equipment manufacturing, GE's primary business. The resolution had also directed the FTC to examine the extent to which those controlling the utility industry were involved in trying "to influence or control public opinion on the question of municipal or public ownership of [electric utilities]," but this was blocked by a ruling from the attorney general.[48]

In addition to using the same sorts of public information as were used in the 1916 report, the FTC also requested and received information directly from high-level holding companies and others thought to be in control of utilities directly or through subholding companies. This information included stock ownership, interlocking directorships, and the financing and profitability of holding company systems.

The report exonerated the industry from possessing the type of monopoly control claimed by Senator Norris. "It is obvious that in 1924, neither the General Electric Co. nor any other single power interest, or group of clearly allied power interests, substantially monopolized or controlled the generation, transmission, and sale of electricity in the United States."[49] The analysis supported this conclusion but also showed the important role of holding companies in the industry. The largest system, Electric Bond & Share's, generated about 12 percent of the nation's electricity. The corresponding percentages for the seven largest holding company systems and all holding company systems were 41 percent and 80 percent, respectively.[50]

Although finding no evidence to support the worst fears of the Senate progressives stated in the resolution, the report did raise a couple of concerns. It provided evidence that since the data had been collected industry concentration had increased. It noted that investors might be deceived about the risk of holding company securities because of the extent of pyramiding and financial leverage (discussed in chapter 6). In addition, increased interstate activity by holding companies was noted as a justification for future congressional attention.[51]

At the time the Senate passed the original resolution, the FTC was in the hands of progressives. Very soon after that, the agency came under the control of conservatives, and it was then that the report was prepared. Progressives had little confidence in the commission or the report.[52] A few months following its issuance, Senator Walsh of Montana called for a new investigation of electric utility holding companies, this time by a Senate committee. After considerable wrangling, the Senate adopted a resolution on February 2, 1928, for an investigation of both industry concentration of control and use of propaganda. The resolution called for a Senate committee to conduct the investigation, but, to the dismay of progressives, it also was assigned to the FTC. This time, however, the FTC was given subpoena power and directed to collect evidence at public hearings from sworn witnesses and to use its own experts to examine company records and physical assets. Reports from these hearings were continuously provided to the Senate and made public. The stock market crash and Great Depression dramatically changed the nation's political climate, and control of the FTC shifted back to progressives. Between 1928 and 1935, ninety-six Senate reports, including transcripts and documents gathered in the hearings, totaled over 65,000 pages. At the investigation's end, three final reports filled over 1,500 pages.

The report provided evidence of an increase in the share of all private generation under the control of holding companies and evidence that control had become much more concentrated (figures 5.4 and 5.5). Control of all private generation by holding company systems increased from 79.5 to 82.3 percent, although inconsistencies between the two reports may have understated that increase.[53] In 1924, the seven largest holding company systems controlled 42.8 percent of all privately owned generation. In 1929, just three systems controlled a larger share. The areas served by holding companies substantially increased between 1925 and 1932. Figures 5.6 and 5.7 show counties where a holding company subsidiary in each year served at least one community. The pattern was sometimes described as a "crazy quilt" because the operating utilities of so many systems were widely dispersed.

Before the investigation was complete, the nation and utility holding companies had experienced the worst of the Great Depression. Between 1928/1929 and 1932, consumer prices fell about 24 percent. The deflation-adjusted value of utility-generated electricity fell by 7.3 percent.[54] Most industries experienced much greater revenue losses. The nominal price for electricity to residential users fell by 14 percent, while that for all services actually increased. The use of both electricity and gas declined, especially by industrial users.[55] Between 1929 and 1936, eighty-two holding companies and fifty-two operating utilities that were in holding-company systems had defaulted on bonds or had gone into bankruptcy or receivership. $641 million owed preferred-stock holders had not been paid.[56] In

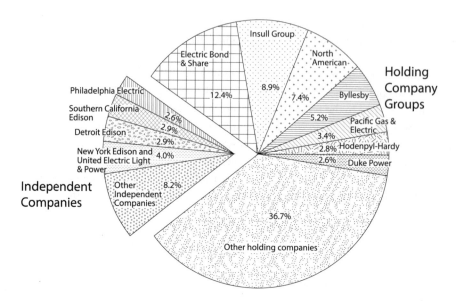

FIGURE 5.4. Percentage of US privately owned utility generation by source, 1924

Source: US Federal Trade Commission 1927, 36–37

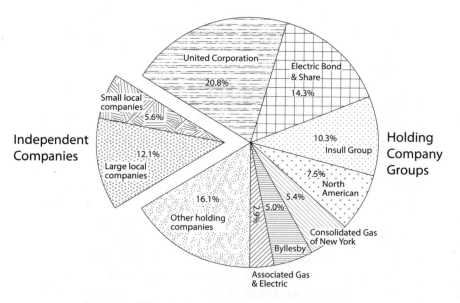

FIGURE 5.5. Percentage of US privately owned utility generation by source, 1929

Source: US Federal Trade Commission 1935a, 38

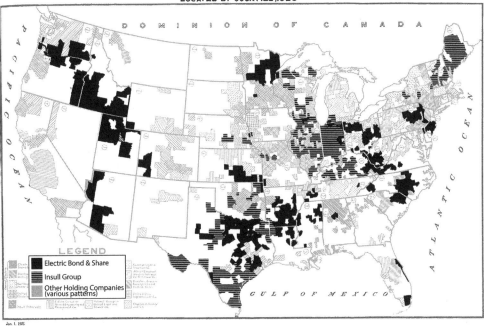

FIGURE 5.6. Counties served by operating companies in holding company systems in 1925

Source: US Federal Trade Commission 1927, opposite 176, and see note 10

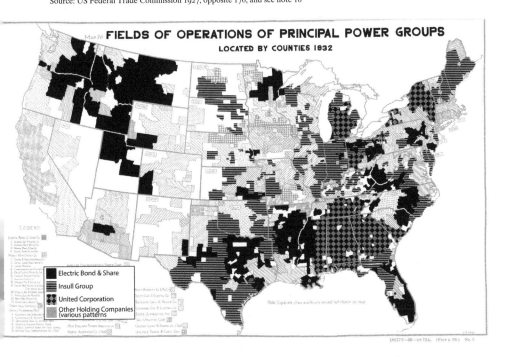

FIGURE 5.7. Counties served by operating companies in holding company systems in 1932

Source: US Federal Trade Commission 1935a, opposite 87, and see note 10

June 1932, the stock market overall had lost 82 percent of its September 1929 value. Holding companies lost 90 percent of their value during the same period. By December 1934, the stock market rebounded with a 122 percent gain, but the holding company index had increased only 4 percent. Over the entire period from September 1929 to December 1934, the index for all stocks lost 59.8 percent of its value, but the loss for holding companies was 89.3 percent. Only two industry indexes fared worse.[57] At the end of 1940, dividend payments had been missed on 56 percent and 28 percent of all preferred stocks issued by holding and operating companies, respectively.[58]

In addition to documenting the increase in industry control by holding companies, the reports also concluded that holding company practices victimized both investors and the users of electricity. Under holding companies, the electric utility industry was also found to have developed a massive and sophisticated public relations (or propaganda) operation that used deceptive techniques to influence public opinion. The report provided a series of alternative recommendations ranging from major reform to dissolution of holding company systems. The Great Depression brought a rapid deterioration in the political influence of privately owned electric utilities. Despite intense effort, those utilities were unable to prevent Congress from passing a "death sentence" that radically reshaped the industry's structure. The next chapter analyses the FTC's findings and discusses Congress's response.

CHAPTER SIX

Public Utility Holding Companies

Indictment and "Death Sentence"

O pponents of public utility holding companies were vindicated by the FTC report. Holding companies had increased and concentrated their control over the industry. They engaged in practices that victimized both investors and electricity users. States and state regulation were largely powerless to control these practices; some state laws actually encouraged their use. The report also revealed the existence of a large and sophisticated organization within the industry that sought, sometimes covertly, to manipulate public opinion, further confirming the suspicions of those opponents. The report also noted positive attributes of holding companies, although it gave little weight to the most important of those when the report was issued, their ability to create large fully integrated networks. In response to the FTC's findings, the federal government instituted radically reactionary changes to the industry's structure that sacrificed the formation of new multistate integrated networks.

The Positive Effects of Holding Companies

The important and positive role holding companies played in the formation of the electric power industry was uncontested. Holding companies likely played a critical role supporting the rapid growth in the industry's capital investment between 1918 and 1924 (figure 4.3), when electrification was having a major impact on American manufacturing and standard of living. Gross investment stopped growing after 1924, as did the need for investment support. Holding companies continued to provide needed technical and managerial support, but they eliminated a more efficient source for that support, consulting companies.

The major positive role of holding companies after 1925 was the continued consolidation of existing operating companies into larger integrated networks, which may have partially explained the decreased need for new investment. As explained in the introduction, and demonstrated by the Superpower experience in chapter 4, it is very difficult to create a fully integrated network from separately owned utilities.[1] The creation of such networks was not the driving force behind the growth of many holding company systems, however. Such networks required control of operating companies with adjacent service areas. By 1924, many holding company systems consisted of widely dispersed operating companies. The helter-skelter scramble for operating companies by the wave of holding companies created in the late 1920s aggravated this situation. However, some holding company systems did create large integrated networks. All of the operating companies in some holding companies, including Pacific Gas & Electric, Consolidated Gas Company of New York, and Duke Power, were adjacent. These companies had little difficulty reorganizing as large single operating companies, each controlling an integrated network. Holding companies had a particular advantage in creating multistate integrated networks. Networks they created still exist today that were and remained important parts of the nation's electric utility industry. Southeastern Power & Light created a modern network spanning areas of Mississippi, Alabama, Georgia, and Florida. Electric Bond & Share created two multistate networks, one that operates in Mississippi and Louisiana and another in portions of Michigan, Indiana, Ohio, Kentucky, West Virginia, Virginia, and Tennessee.[2] In 1933, 94 percent of all interstate electricity transmission was within the same holding company system.[3] In 1952, sixteen multistate networks created by holding companies collectively accounted for 23 percent of all investment and revenue in the US electric utility industry.[4] The creation of new networks stopped when holding companies were eliminated.

The Negative Effects

The FTC report identified ways holding companies thwarted state utility and securities regulation and income tax collections. They concealed the true risk of their securities acquired by millions of investors, many of whom endured losses. The fury created by those losses conditioned many to believe they were the victims of actions that were immoral, unethical, and even criminal by individuals that used them to amass great wealth. The FTC investigators shared those feelings, and the tone of the report was an indictment of holding companies:

> It is not easy to choose words which will adequately characterize various ethical aspects of the situation without an appearance of undue severity. Nevertheless the use of words such as fraud, deceit, misrepresentation, dishonesty, breach of trust, and oppression are the only suitable terms to apply if one seeks to form an ethical judgment on many practices which have taken sums beyond calculation from the rate paying and investing public.[5]

The steady stream of findings released during the course of the investigation attracted considerable press attention, and several books repackaged those findings for a popular audience. This attention both reflected the public's interest and contributed to its outrage.[6]

Did those findings justify the radical action that reshaped the electric power industry in ways that lasted for decades? Was there something in the nature of a public utility holding company that encouraged and enabled it to victimize its investors and customers more than did other companies? The answer to the latter question is a qualified "yes."

Exploiting Investors

Holding companies were accused by the FTC of victimizing investors by misrepresenting the true net worth and profits of companies that issued securities and the risk of holding those securities. Net worth and profits measure the financial strength of a company. Misrepresentation of these could have led investors to have overoptimistic expectations about future earnings, creating unjustifiably high common stock prices and making other securities seem safer than they really were. The lower perceived risk would have reduced the returns bonds and preferred stocks would have needed to pay investors. Potential and current investors relied on the accountants who audited the firms to provide accurate and unbiased reports. That reliance was often unjustified. Deficiencies in audit reports were exacerbated by a lack of uniform standards for audits. Some of the practices of holding companies were reprehensible, and reform was fully justified.

Problems with the information provided outside investors were not limited to public utility holding companies. A Senate committee (commonly called the Pecora Commission) conducted hearings between 1932 and 1934 on the sale of securities to the public and released a sequence of twenty transcripts and a final report.[7] The investigation did include public utility holding companies (one hearing was devoted to Insull), but it dealt primarily with investment banks. Its report, like that of the FTC, had a strong tone of outrage. Although the Pecora Commission identified some of the dubious practices by holding companies later reported by the FTC, its final report only briefly mentioned public utility holding companies. Railroad holding companies were the primary examples of holding company abuses in that report. The findings of the Pecora Commission spurred the passage of the first federal securities law, the 1933 Federal Securities Act, which was specifically designed to require corporations to provide better information to outside investors. The following year Congress extended these requirements and created the Securities and Exchange Commission (SEC) to regulate the sales of securities.[8] The SEC directed the accounting profession to develop uniform auditing standards, which it began in 1939.

Were public utility holding companies more deceptive than were other companies? The ideological opponents of private utility ownership could have been expected to hold that view, but others held it as well. For example, the president of the Investment Bankers Association of America noted in 1927 that corporations in general had improved the information they made available when issuing securities, but specifically excluded public utility holding companies from this praise.[9] The multilayer organizational structure of holding company systems could both create and conceal financial fragility and the riskiness of their securities. Although multilayer organization may have been present in other industries, the rapid acquisitions by public utility holding companies of other holding companies likely made this especially common. Utility holding companies were unique in having subsidiaries subject to state regulation. Control of regulated utilities enabled types of deceptive behavior that would otherwise have been unavailable. That behavior, however, was used more to exploit customers than investors.

In many states, regulated utilities had to have the prior approval of the regulatory commission before issuing securities. This gave regulatory commissions power to prevent some dubious financial practices. Although operating utilities were subject to state regulation, holding companies were not. Their securities might have been subject to "blue sky" laws used by some states to protect those investing in any industry from fraudulent securities offerings. These laws generally required offerings of new securities to receive prior approval by a state agency and to pay large fees, but they did not effectively protect investors. Some state laws were inadequate, and some states were lax in the enforcement of the laws they did have.[10] Some states exempted certain types of bonds or securities sold on major exchanges from their blue sky laws. Corporations sidestepped them, even for legitimate securities, by using mail to sell them from outside the state

The final FTC reports devoted considerable space to criticizing certain accounting practices used by holding companies and describing preferable alternatives. By modern standards, many of those practices are outrageous and illegal. They were not illegal at the time, however, and some were used by firms outside the utilities industry. By the standards of the time, the FTC's outrage may have been unfair. A deficiency of the FTC's report is that it did not explain exactly how investors were harmed by the practices it criticized. Some of those practices are widely used today and are not regarded as harmful to investors, raising questions about the FTC's conclusions. To judge the accuracy of the FTC's criticisms, the actual or potential harm done to investors must be determined. This is made especially difficult because many of the accounting concepts used by the FTC are archaic and no longer used. Perhaps because of these difficulties, there have been no thorough modern efforts to analyze the FTC's findings.[11] There is no doubt, however, that practices used by holding companies would have provided investors with distorted information that concealed negative aspects about a company's financial situation. However, investors were aware of the use of many of them, although that awareness itself would not have enabled a correction of the distortions.

A firm's net worth is the values of its assets minus liabilities at one point in time. A firm's profit is the difference between its revenue and operating expenses across a period of time, usually a year. A balance sheet, showing the firm's net worth, and a profit and loss statement, showing its profits, were issued following an audit. For a common stockholder, net worth measures the current equity owned collectively by all common stockholders. Expected future profits, not current profits, are a primary determinant of a stock's market value. Since it is impossible for that to be objectively determined, a prospective investor in common stock must combine the firm's current financial situation with other information to form an opinion about future profits. A firm with a large net worth was less likely to go bankrupt. If it were liquidated, net worth provided a source of funds to pay other investors if the sale of assets alone was insufficient. High net worth was an indication that the firm's securities were a safe investment. Large profits provided a source of funds to pay income to securities holders and were another indicator of safety.

The managers of a company possessed the most information about both its current net worth and its current profits. They, however, had an incentive for both values to be as high as possible. Higher values for these measures would increase the value of the company's securities, possibly increasing their personal wealth, and would make it easier to sell additional securities. The greater the perceived safety of securities, the lower the return necessary to attract investors. Those lower payments to investors would increase net value and profits. The incentives to increase a firm's net value and profits encouraged managers to behave in the best interests of investors, but they also made those managers an unreliable source of objective measures of the firm's financial strength.

To achieve as much objectivity as possible, audits and reports should have been done by outside accountants who had no incentive to provide biased information. The FTC found this was not always the case. Ideally, audits would have been done using the same procedures in all audits in all industries by all accountants, any of whom would have obtained the same results, comparable across industries. Not only did a lack of standards prevent this, but also the determination of some values inescapably required a degree of subjective judgment. Today, investors can often also get information from independent analysts, which can provide a check on the company's reports. Such information was uncommon in the 1930s and earlier.

Determining the value of a company requires assigning values to its assets. There are two categories of assets: tangible and intangible. Tangible assets include land and capital equipment. Intangible assets include items without physical existence but which contribute to the company's earning power, including franchises, rights (such as a right granted to use a hydropower site), and other privileges. Valuing intangible assets was more difficult, but considerable controversy also existed over how tangible assets were valued.

The accounting concept used by the FTC closest to measuring a company's net worth was "surplus." It purported to measure how much would be left over if a company were liquidated and all of its obligations paid, including the amounts required to redeem preferred stock. The FTC maintained that surplus should have been divided into two classes: earned surplus and capital surplus. Earned surplus contained only amounts the company had received in profits but not distributed to common stock-holders, including profits from operations and gains from the sale of as-sets. Capital surplus contained all other net assets, including the funds the corporation received from the original sale of common stock and the value of stock that was surrendered or donated to the company. The FTC's posi-tion was that capital surplus was to be retained by the corporation unless it was liquidated and could not be paid as dividends. Only earned surplus should have been at the disposal of the board of directors.[12] Most holding companies, however, were incorporated in states that allowed the payment of dividends from capital surplus, and the number of such states allowing this practice was increasing.[13]

Holding company systems commonly included in surplus items that compromised its usefulness to investors and that the FTC regarded as bogus. Again, these practices were used in states in which they either were not prohibited or were explicitly permitted by incorporation laws. An ex-ample of such a practice was the inclusion of the value of stock dividends in surplus. A stock dividend (equivalent to a split) occurred if a corpora-tion issued new stock and distributed it to existing stockholders in fixed proportion to the stock they already held. Increasing the total number of shares of a company's stock would have diluted the value of each stock and left the total value of all stocks unchanged. The new stock could be sold, but the receipt from that sale was not income but came from the value of existing stock. The FTC's criticism was correct, but the practice was legal in some states.

The largest source of improper amounts credited to surplus came from the overvaluation of an operating utility's assets. In addition to being a source of investor deception, if these overvaluations were accepted by state regulators as part of the rate base, they also resulted in higher rates, exploiting electricity users. The FTC's position was that the basis for valu-ing tangible assets in the company's books should have been their origi-nal cost and that increases in an asset's value before it was sold should not have been included in surplus. Neither of these was standard industry practice, and this resulted in what was probably the greatest abuse of both investors and customers by holding companies.

Write-Ups

A "write-up" occurred when the management of a company increased the value of an asset on the firm's books although the actual physical characteristics of that asset were unchanged. This could occur when the asset legally passed from one subsidiary to another, because of either an intrasystem transaction or a corporate reorganization. Increasing the book value of an operating company's assets allowed a holding company to report a higher value of the securities of the operating company it held. In a multilevel system, this could ratchet up the reported net worth of companies at every level, increasing the apparent safety of all their securities. Arguably, if the operations of two operating companies were merged, the profits of the new company might exceed the combined profits of the two, and the new company would then be more valuable than the two it consolidated. At best, however, those higher profits would come in the future, and immediately assigning a higher value to the combined company was speculative. In many cases, the magnitude of a write-up was so great that no claim that it was based on the economies of consolidation was credible.

The greatest abuse occurred when a write-up was "capitalized," that is, when it was used to support issuing more securities. If that happened, the system took on a new obligation to pay returns to investors of the new securities even though there was no actual change in the system's physical or financial situations. The new securities increased the risk of existing securities, but the increased reported surplus concealed this. The derisive term for securities issued based on write-ups was "water," implying they had no real value. Although the new securities harmed existing investors by deceptively increasing the risk of their securities, they could provide a large cash benefit to the system's top common shareholders.

Here is a simplified example of how a write-up could work. Suppose one company in a holding-company system sold an operating utility to another company in the same system at a higher price than was originally paid. The buying company then financed the purchase by issuing and selling new securities and giving the funds to the selling company. The reported surplus of the acquiring company was unchanged: the obligations to the new securities holders exactly offset the value of the new asset (the acquired company). The selling company reported a profit from the increased sales price, which increased its earned surplus. Thus, the entire system's reported surplus increased, although there was no actual physical change in any of its assets. The selling company then distributed the profit it received in dividends to its common stockholders. In essence, the receipts from new securities were paid directly to common stockholders. They benefitted, but the securities of all investors became riskier. At best, capitalizing the write-up and distributing it as dividends shifted the (possible) future profits to the system's present top stockholders. Following the nation's economic collapse, write-ups increased the losses of outside investors.

An example of a write-up occurred in 1924. Cities Service, then a top-level holding company, acquired the majority of the stock of an operating company, Colorado Power, for slightly more than $3 million, in an arms-length transaction with unaffiliated parties including the Electric Bond & Share system. A few weeks later, Cities Service sold the stock to an existing operating company subsidiary, Public Service Company of Colorado, for slightly more than $9 million. Public Service paid for the stock by issuing $9 million in bonds. The two companies merged, and the new company's value included the $6 million write-up. The physical assets of the merged company were the same as those of the two companies it replaced, but its value was $6 million greater. The stockholders of Cities Service made a large profit without having had to put up any money, but the system's total debt increased. Any contention that expected higher future profits justified a tripling in the value of an asset within a few weeks seemed questionable. There were numerous other cases of write-ups arising from transactions between firms in the same holding company system.[14] The FTC found that write-ups in eighteen holding-company systems constituted over 22 percent of the total value of the operating utilities in the system. The total value of those write-ups exceeded $1.4 billion (between $17 and $24 billion in 2014 dollars, depending on when the write-ups were made).[15]

Write-ups were possible because the common basis for the valuation of tangible property was "replacement cost new," what it would have cost at the time of the write-up to duplicate the physical assets acquired based on an engineering appraisal. A truly objective appraisal would have been difficult, but the FTC found that the appraisals were frequently made by employees of the same holding company or by others dependent on the system.[16]

Depreciation

As a physical asset ages, it wears out or becomes obsolete, and its value decreases. Depreciation as an accounting concept seeks to measure the amount of this decrease in value that occurs in each year of the asset's life. A prediction is made of the life of a new piece of equipment, and an assumption is made about the temporal pattern of the decline in its value.

For example, straight-line depreciation assumes each annual decline is the same. The predicted life and assumed pattern of deterioration are used to assign a value to each year's depreciation. That amount is treated as an operating expense and subtracted from the year's revenue in determining profit, and it is subtracted from the total value of the physical capital on the firm's books. Ideally, when a piece of equipment is taken out of service, the sum of each year's depreciation will equal the equipment's original cost. Of course, the predicted life may turn out to have wrong, but that experience is taken into account in predicting the life of other capital equipment to make average expected life of all capital equipment as accurate as possible. Sometimes depreciation is credited to a sinking fund, which can be used to purchase replacement equipment. Today depreciation is the accounting technique universally used, and it was the method used by most industries at the time of the FTC investigation. It was not, however, the method generally used by the electric utility industry.

Instead of using depreciation accounting, utilities used a method termed "retirement" accounting. No adjustment was made to profits or net value during the life of the equipment. Instead, all of the equipment's original cost was included in operating expenses and deducted from the firm's books in the year it was removed from service. The book value of a utility's physical assets thus did not account for the deterioration of equipment still in use. When gross investment was increasing (as during most of the life of the industry) reported profits and net value using retirement accounting would have always been higher than under depreciation accounting, inflating the value of reported surplus.

Under retirement accounting, the utility created a retirement account into which profits would be placed at the discretion of the board of directors. The funds in that account could then be used to purchase new capital equipment, similar to a sinking fund. The purpose of a retirement account was not, however, to pay for capital equipment. Its primary purpose was to reduce variations in common stock dividends. If the directors judged a year's profits to be unusually high, rather than increasing dividends, they could add to the retirement fund. If, however, the directors judged a year's profits to be unusually low, or when a particularly expensive piece of capital equipment was purchased, they could take funds out of the retirement account to pay that year's dividends. Advocates of retirement accounting argued that dividends should be determined by the usual level of earnings and not be affected by year-to-year fluctuations. It was entirely up to the judgment of the directors as to what constituted the "usual" level of earnings. With rare exceptions, the retirement account was never sufficient to cover the cost of replacing capital equipment.[17]

When a utility's financial situation deteriorated, retirement accounting provided ways to conceal that from investors. When capital equipment was taken out of service, some utilities failed to include its costs in expenses or to deduct its value from those of its total assets.[18] Such behavior might have been expected after the 1929 crash, but it happened earlier. In 1926, for example, Florida Power & Light's profits and reserves were insufficient to cover the cost of retired equipment, so it kept over $1 million worth of that equipment on its balance sheet. Between 1926 and 1929, a number of holding company profits had fallen to a point that depreciation accounting would have eliminated dividend payments to preferred stockholders. With the use of retirement accounting, those companies were able to pay dividends not only to preferred stockholders but to common stockholders as well.[19]

Although retirement accounting would find no support from accountants today, it did then. Its use was approved by state regulators and included in the Uniform Classification of Accounts developed by state regulators' national association (with the active participation of the NELA).[20] By enabling a higher valuation of a regulated utility's assets (rate base), retirement accounting increased allowed revenue and electricity prices.[21] If, despite using retirement accounting, a utility found it in its interest to add a depreciation charge to its operating expenses, regulators often gave their approval. The standard used by regulators in such a situation was "reasonableness," rather than depreciation accounting's objectively determined amount.[22] This gave utility management latitude in determining the timing of the impact equipment wear and tear had on allowable revenue.

The FTC found other methods used by holding companies to victimize outside investors. A number of holding companies tried to manipulate the market price of their stocks. The FTC maintained that often the intention of the manipulation was artificially to raise the price of a stock before the issuance of new shares, particularly when the new stocks were offered by subscription to existing stockholders. Once the need to support the stock price was over, selling those surplus shares would depress stock prices, possibly resulting in losses to investors. In some instances, a company or its closely aligned interests purchased and sold over 90 percent of the total monthly volume of a company's stock traded on a public exchange. This type of manipulation was risky to the company because to raise prices, it had to buy more shares than it sold. The FTC provided considerable detail on stock market manipulation by the Insull interests when the system was refinanced shortly before the crash. In that case, the FTC maintained that the Insull system had created new companies with new stock issues to absorb the surplus manipulated stocks.[23] After the 1929 stock market crash and continuing decline in stock prices, efforts at manipulation generally created losses. Stock market manipulation was a problem in industries other than public utility holding companies. As the FTC noted, the Pecora Commission found that financial firms not connected to the utility industry had also frequently engaged in securities-price manipulation.[24]

The FTC documented and adamantly condemned the use of stock without par value by utility holding companies. Par value originally represented the amount of money put in the corporation by each share of stock when it was originally issued. If the original stockholder paid less than its par value, he or she could be held personally liable for the difference. It was common, however, for the original price of stock to exceed its par value. The subsequent market value of the stock was unaffected by its par value. Par value is now generally regarded as meaningless, and no-par stock is commonly used and not viewed as suspicious. The report did not explain how investors were victimized by no-par stock. It is difficult to determine if the FTC was simply mistaken about the issue or if some unexplained factor no longer in existence justified its position.

Why Use Deceptive Accounting?

At best, retirement accounting and unjustified write-ups that did not affect regulated rates shifted profits from the future to the present. Because future profits cannot be known with confidence, such an action is inherently risky. The reason this might have been done by the unscrupulous is obvious, but why would a reputable company that expected to remain in business mortgage its future? For such action to make sense, those responsible must have had a strong reason to believe that profits now were far more valuable than future profits. There are plausible explanations for why that belief was held. Prior to 1915, utilities needed each year to raise more money from investors than they received in revenue (figure 3.2). For the first half of the 1920s, gross investment averaged 64 percent of total revenue. Shifting profits to the present may have been seen as necessary to induce the level of investment needed to feed the industry's appetite.

Continual rapid growth may have made shifting liabilities to the future not seem very risky.

By the second half of the 1920s, the need for new funds to finance capital equipment had declined. This was, however, the period when holding companies were scrambling to acquire operating companies and other holding companies at increasing prices, and control of the industry was falling in the hands of fewer and fewer companies. For holding companies whose managers wanted to maintain some control of the industry, acquiring control over more operating companies was essential. For holding company managers without such lofty goals, growth could still be seen as an imperative. The prices paid for control of operating companies was increasing. The stockholders of a holding company acquired by another could receive larger capital gains if the purchase included more operating companies. Eventually one, or a few, holding companies would be left standing in control of the industry and further acquisitions would cease. During a brief window of opportunity, funds from investors now rather than later promised high rewards justifying some additional risk. Such an attitude could also help explain why some systems, notably the one controlled by Insull, were anxious to acquire control of operating utilities whose locations offered little or no opportunity for integration. In the late 1920s, the economy seemed to be on a path of endless growth, and the electric utility industry was riding the crest of that growth. This environment did not encourage a conservative approach to business decisions.

When a holding company system faced the imminent prospect of failure, those in charge may have resorted to "gambling for resurrection," engaging in risky behavior that had a slight chance of saving the business but was more likely to result in even greater losses.[25] This could have included misrepresenting the company's financial situation so that investors continued to purchase securities. The misrepresentation may have been done with the hope that conditions would improve soon enough that the new investor obligations (along with the old) could be met. Part of the motivation for this may have come from the fact that a business's managers and common stockholders were the first to lose, and a collapse could result in their losing everything. They had little or nothing more to lose if total losses increased further. When the collapse occurred, gambling for resurrection would have contributed to investor anger and been seen as fraud. Although eventually acquitted, the FTC accused Insull, when his system appeared near collapse, of having accelerated sales of securities to customers, relied on fictitious profits, and manipulated the stock market.

Other holding company executives were convicted and sent to prison, including officers of the W. B. Foshay holding company, whose failure became apparent even before the economic collapse. It had long advertised to investors, "The company has never failed to pay dividends—it pays them monthly; it meets its obligations to shareholders on time—all the time." Its profits fell to zero and became negative even before the economic collapse. Nevertheless, it kept its word and continuously paid dividends through October 29. The next day, barely a week after the stock market crash, it declared bankruptcy.[26] Gambling for resurrection may partly explain the FTC's condemnation of the ethics of utility holding company executives. However, before such behavior could occur, a holding company system must have been on the brink of collapse. The rapid acquisition of holding companies by other holding companies contributed to a fragile financial structure that made even a slight downturn create such a situation.

Pyramiding and Leverage

Each time a holding company acquired another, it added a new layer to the existing system. A diagram of such a system was like a pyramid—one company at the top and ever-increasing numbers of companies at lower levels. The FTC objected to two characteristics this structure enabled: excessive financial leverage and control of a system by individuals whose personal investment was a tiny fraction of the total. Concern over excessive financial leverage was justified. Concern over the second characteristic might also have been a legitimate cause for concern, but that condition is extremely widespread today and is not now a source of alarm.

Leverage occurs whenever an investment is partially funded by the sale of financial instruments that pay a fixed return, including bonds and preferred stock. The higher the proportion of the total funded by fixed-return investments, the greater the leverage. Some degree of leverage is normal in all businesses, but overreliance on fixed-income securities created excessive leverage in some holding company systems. With leverage, changes in the return to the investment in an operating utility multiplied the change in the return to the system's top stockholders. A small increase in an operating company's return caused a huge increase in that of top stockholders. However, even if an operating company remained profitable, a small decrease in its return might not have just reduced the return to top stockholders, it could have bankrupted one or more of the system's holding companies. Fear of financial fragility was the source of the rule of thumb that no more than 50 percent of a company's investment funds should come from bonds. Multiple layers in a holding company system could have made the system's reliance on fixed-system securities less apparent to investors than it actually was.

The original design of some holding company systems often included a few layers. In the original structure of Electric Bond & Share (chapter 5), for example, operating companies were owned by subholding companies controlled by the top holding company. When a holding company acquired another, there may have been an intention to merge layers simplifying the system's structure, but this took time. In the meantime, acquisitions increased. As the number of layers increased, so did financial leverage. Figure 5.2 shows the layers between top stockholders in the Insull system and one of its operating companies. When holding company systems with different numbers of layers were acquired, the number of layers between the acquiring company and an operating company varied. Most of the operating companies in the Associated Gas and Electric system were separated from the top company by three layers of subholding companies (five layers in total). Two operating companies in the system, however, had five layers of subholding companies between them and the top company, seven layers in all.[27] The total number of legally distinct corporations in some holding company systems was in the hundreds.[28]

Figure 6.1 illustrates leverage in a hypothetical holding company system with only three layers and illustrates the pressure for operating companies to pay unjustified dividends. System diagrams usually pictured the operating companies at the bottom and the top holding company at the top. For clarity, this one is inverted. The diagram simplifies aspects of actual systems that are unnecessary to illustrate leverage. Each layer has only one subsidiary company, but in actual systems, all holding and subholding companies had multiple subsidiaries. Each company shown is financed 50 percent by bonds and 50 percent by common stocks. The bonds issued by all layers pay the same interest rate, five percent. All of the common stock in the system is owned either by a holding company or by the top stockholders. There are no preferred stocks. Since preferred stocks paid a fixed return, their effect on leverage was the same as bonds, although, unlike bonds, they could not force bankruptcy. The two holding companies have no operating expenses. Like actual holding company systems, all funds raised from the sale of securities by all companies were used to

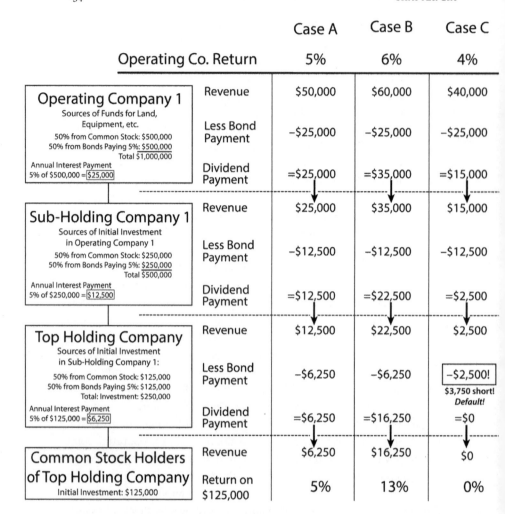

FIGURE 6.1. Effects of financial leverage in a holding company system

Source: Prepared by author

purchase the capital equipment of the operating company, which was the sole source of all profits.

The total cost of the illustrated operating company's capital equipment is $1 million. The operating company raised half that amount by issuing and selling bonds and half by selling common stock to the subholding company. The subholding company financed half the $500,000 it paid for the operating company's common stock by selling $250,000 of its own bonds and $250,000 worth of stock to the top holding company. The top holding company, in turn, sold bonds for half the amount and obtained the other half from the system's top stockholders.

Column A shows the flow of money if the net revenue to the operating company was $50,000, a 5 percent return on its $1 million worth of capital equipment. The $500,000 it raised from the sale of 5-percent bonds required an interest payment of $25,000. That interest is paid, and the remaining $25,000 is paid as a dividend to the subholding company. The subholding company pays the $12,500 in interest owed its bondholders and pays the remaining $12,500 as a dividend to the top holding company. That company pays its bondholders the $6,250 in interest they are owed, and the remaining $6,250 is paid as a dividend to the system's top stockholders. Their original investment was $125,000. The $6,250 dividend is a 5 percent return on their investment.

In column B, the holding company's profits have risen to $60,000, a 6 percent return on its $1 million of capital equipment. This does not change the amount owed bondholders; they are paid the same $25,000 as before. The remaining $35,000 is paid to the subholding company, which pays its bondholders $12,500 and pays $22,500 in dividends to the top holding company. The top holding company pays the interest it owes its bondholders, and the remaining $16,250 is paid as a dividend to the system's top stockholders. Their dividend is a 13 percent return on their $125,000 investment. A 1-percentage-point increase in the return to the operating company's investment causes an 8-percentage-point increase in the return to the top stockholders.

Column C shows the system's fragility. Net revenue to the operating company falls to $40,000, a return on its investment of 4 percent, a 1-percentage-point fall from case A. The operating company is still profitable. After bondholders are paid, the subholding company receives $15,000 in dividends and the top holding company receives $2,500. However, the top holding company owes its bondholders $6,250 in interest, $3,750 more that it received. It can only make a partial payment to its bondholders and defaults on the rest. The top stockholders receive nothing. A continuing return of only 4 percent to the operating company will eventually bankrupt the top holding company.

The total interest payments owned all of the system's bondholders equal $43,750. As long as the operating company pays a dividend of at least this amount, bankruptcy will be avoided. If there were more layers, each with the same financing, leverage would have been increased, and an even larger dividend payment from the operating company would have been necessary to keep the entire system afloat. Although each company is financed only 50 percent by bonds, the relevant leverage to a bondholder is the total percentage of the operating company's $1 million financed by bonds at each level from the company issuing the bonds through the operating company. For bondholders of the top holding company, the relevant percentage was 87.5. Such information was not easily obtained.

To avoid bankruptcy, the top stockholders might gamble for resurrection by forcing the operating company to pay dividends above its actual profits. This could be done only by reducing that company's capital below $1 million, perhaps by selling assets. To earn the minimum $47,500 to avoid default, its rate of return on the reduced capital would have to increase. If this did not happen, total losses to bondholders would be greater than if bankruptcy had occurred earlier. Gambling for resurrection shifted risk from the top stockholders to the system's existing bondholders. The FTC had little difficulty finding cases of operating companies paying unjustifiable dividends that jeopardized their future.[29]

The FTC was also bothered by the way multiple levels in a holding company system could enable top stockholders to control a total investment having personally made only a tiny fraction. In the example shown in figure 6.1, the operating company's $1 million investment was completely controlled by the top stockholders, whose own investment of $125,000 was only 12.5 percent of the total. In the case of the segment of the Insull system shown in figure 5.2, the FTC calculated that the personal investment of the system's top stockholders was less than 0.03 percent of the total. The FTC saw this as a problem encouraging risky behavior:

> It seems very unsafe to have any form of pyramiding which has such a financial basis, not only on account of the excessive concentration of control over immense masses of property but also because of the opportunity it offers to financial adventures to have too much influence over the general economic interests of the company.[30]

Those who lost money in the collapse of holding company systems were apparently shocked to learn about the small investments made by those who controlled the systems and their small share of the total losses. However, the investment required by an electric utility would have made it impossible for those in control of a large holding company system personally to provide a significant proportion of the total. This situation exists with any giant corporation, regardless of the number of layers, and is the norm among large public corporations today. The common situation today is for a corporation to be controlled by its professional managers, whose personal investment is an even smaller proportion of the total than that identified by the FTC. Considerable research has shown that these managers operate in their own best interests even when that does not coincide with the interests of all stockholders.[31] At the time of the FTC investigation, it was already known that control by management existed in the case of AT&T.[32]

Modern control by managers occurs when the corporation has many stockholders, each of whom owns such a tiny proportion of the total that he or she has no hope of influencing the election of a director. When an election for the board of directors occurs, the company's management will issue a recommended slate of candidates. The largest stockholders, typically institutions such as mutual funds or pension funds, tend to vote for that slate unless they are very unhappy with the current management. Management will solicit the votes of all stockholders, many of whom will not bother to vote. If a director acts against the interests of the managers, he or she will not be on their recommended slate in the future and will not remain a director. Rather than the directors controlling management, management effectively controls the directors. All stockholders, including institutions, will usually own stock in many corporations, making it impossible to become fully informed about the operations of any single one. An investor unhappy with the performance of a corporation will simply sell his or her shares. There are some limits to the management's power. Sometimes a corporate raider or other current or potential stockholder will believe a company's management is doing such a poor job of serving shareholder interests that a campaign will be made to get control of the votes of other shareholders. These undertakings are difficult, expensive, risky, and rare.

Holding companies provided their investors spectacular returns in the 1920s. Even with better information, many investors would likely not have refrained from continuing to invest in holding companies and instead followed a conservative approach that would have better protected them when the nation's economy suffered the largest and longest financial downturn in its history. The anger experienced by the enormous number of individuals who had invested in utilities was, nevertheless, unsurprising. Reform was clearly needed, and prosecution of some holding company system officials was warranted. Some holding company systems, however, were not in the same category and warranted little reproach, especially by the standards of the time. No other industry was investigated to the same extent as public utility holding companies. That absence makes it impossible to compare the behavior of the electric utility industry with that of other contemporary industries.

Exploiting Utility Customers

The FTC made a strong case that holding companies were able to take advantage of electricity users by exploiting state regulation. The objective of state regulation was to enable electricity users to gain the advantages of being supplied by a monopolist without paying rates that included monopoly profits. Many contemporary observers doubted the effectiveness of state regulation, and the FTC seemed largely to agree:

> ... comparatively few states have adopted any thorough-going policy or system of regulation. Those which have done so have encountered insuperable difficulties growing out of constitutional limitations, economic developments, and the attitude of the industry and the courts.[33]

FTC found evidence of a too-close relationship between state regulators and private utilities. Private utilities interests had paid off deficits incurred by state regulators' national association, NARUC. Utilities and utility organizations gave gifts to state regulators, and paid them to address conventions of utility employees. NARUC's position on the superiority of private ownership of utilities over public ownership mirrored that of the private utilities, and the NELA distributed reports and statements from NARUC expressing that position.[34] The NELA's position on federal regulation, particularly as a threat to state commissions, was in lockstep with that of NARUC.[35] Although opposed to federal regulation, the NELA depicted state regulation as tough, vigilant, and effective.

While acknowledging that there was considerable variation among the states in the effectiveness of regulation, the FTC found much to criticize. It faulted some states for adopting laws and policies that facilitated the victimization of investors and utility customers. It found that some state regulatory agencies had employed policies that favored the interests of the regulated utilities over those of its customers. The FTC questioned whether effective regulation was even possible at the state level.[36] State regulators bore responsibility for the use by the industry of retirement accounting to the disadvantage of both investors and the users of electricity. However, some of the most potent methods by which electric utilities could undermine state regulation could not have been prevented by even competent and conscientious state regulators.

Under the regulatory system that had been created in the United States, an electric utility was entitled to receive revenue sufficient to cover its operating costs plus a fair return on the value of its capital, its rate base. When a commission decided to consider changing a utility's rates, a hearing was held in which detailed data from a previous period (often the previous year) was presented by the utility. Based on these data, the commission determined the values of three numbers: the amount prudently spent by the utility on operations (operating costs), the total value of its capital equipment (rate base), and the fair rate of return. The utility was then allowed to charge rates that, had they been used during the period covered by the collected data, would have provided the utility with the target revenue, assuming use of electricity by all customers was unchanged. Chapter 3 discusses a number of flaws inherent in this system. Holding companies widely employed methods that made it difficult for state regulators to make an accurate determination of the value of either operating costs or the rate base.

Operating Costs

If a utility purchased electricity at wholesale from another utility, that cost would have been part of its operating expenses. Since 1927, interstate wholesale electricity rates were unregulated. Operating companies in one state frequently sold electricity to other operating companies in other states that were in the same holding company system. This gave a holding company the opportunity to enable increased rates by an operating company by charging it an excessive price for electricity from an out-of-state affiliate. Surprisingly, the FTC found this had not happened.[37] Holding companies, however, had other effective ways of manipulating operating costs.

Many holding companies were exclusive suppliers of services and construction to their operating company subsidiaries. Except for construction, which affected the rate base, the prices charged for these services were an operating expense of the utility receiving them. Had the operating company provided those services to itself, the regulatory commission could have determined their costs and used that to determine total operating costs. If the services had been provided by competing consulting firms, competition would have made their fees reasonable. However, as noted in the history of Stone & Webster (chapter 5), the holding company system eliminated that alternative. Holding companies were not regulated, and a commission did not have the power to collect the information needed to determine what the services actually cost them to provide. As long as those fees seemed "reasonable," regulators had no choice but to accept them as an operating expense that could be recovered from the utility's customers.

The FTC found that excessive service fees were the norm. Electric Bond & Share, one of the first holding companies to provide services to its operating utilities, had always been secretive about its servicing business, never releasing information about its revenue or its costs. When the FTC requested that information, Electric Bond & Share refused to provide it. When the FTC issued a subpoena, the company challenged it in court, supplying the information only after the challenge failed. Those data revealed that fees for services amounted to 30 percent of Electric Bond & Share's gross revenue. The holding company charged its subsidiaries more than twice the cost of providing them. Other holding companies were even less restrained in their fees. One holding company was found to have charged fees over five times their costs. Abuses went beyond excessive charges. In one instance, a holding company acquired an operating company and then charged it service fees for the period before its acquisition. The fees charged operating companies in six holding-company systems were between 1.4 and 4.2 percent of those companies' operating costs. The FTC did not seem puzzled that holding companies did not charge even higher fees. Excessive service fees were so clearly abusive that several holding company systems quickly lowered them or adopted methods preventing them from becoming excessive.[38]

The Rate Base

The larger the value of the rate base, the higher the rates regulators would approve. The Supreme Court in *Smyth v Ames* made impossible an agreement on the correct procedure to value a utility's existing capital equipment (chapter 3). As with operating costs, the FTC criticized some practices used in determining the rate base, including "reproduction cost" and the inclusion of "going value," but their use did not depend on the existence of holding companies, and they had been upheld by courts.[39] When a holding company constructed a facility for an operating company subsidiary, the price charged by the holding company, not the cost of construction, was the value included in the rate base.[40]

The biggest issue involving holding company manipulation of the rate base concerned write-ups. In principle, a write-up of a regulated utility should not have affected its rate base. When one utility purchased another, both utilities would already have had a rate base, and their sum should have been the rate base of the new utility. Since the rate base was supposed to be the value of the utility's physical capital, industry spokespersons argued that since a write-up did not affect that physical capital, it would affect neither the rate base nor a utility's rates.[41] Despite the logic of this argument, the FTC found that write-ups were commonly added to the rate base.

Even if inclined to disallow inappropriate write-ups, regulators did allow them, sometimes reluctantly. Excessive write-ups financed by new securities required payment of additional interest and preferred-stock dividends. If write-ups were not included in the utility's rate base, it could have put the operating utility in financial distress. That distress would have adversely affected the utility's ability to attract new investment funds required for actual physical equipment. A utility in financial difficulty was likely to scrimp on maintenance, causing the quality of service to deteriorate perhaps to an unacceptable level. If bankruptcy occurred, management upheavals affecting reliability and quality of service were a distinct possibility. Financial distress by a monopoly utility had far greater impact on customers than such distress by a company in a competitive market, because customers could not switch to a healthy competitor. The total value of electricity far exceeded its cost, and reliability was a major component of that value. The cost to electricity users of higher prices might well have been less than that of a loss of reliability. Regulators admitted to allowing write-ups for this reason even when they knew they were being manipulated.[42] In addition, when a large number of state residents owned utility securities, disallowance of write-ups would have threatened their investments. A regulatory commission that allowed this to happen could have found itself losing popular and political support.

The FTC found that, with few exceptions, regulatory commissions with clear authority to exclude write-ups from the rate base had not exercised that authority. In other states, regulatory commissions simply lacked that authority.[43] Many examples were given of write-ups entering the rate base, including the write-up involving Cities Service and Public Service of Colorado described previously. If a state commission approved the purchase of one operating company by another, it might have had no control over the price that was paid. Approval of the purchase was likely to result in approval of the securities required to complete the purchase. A report issued by an investigative committee of the Pennsylvania legislature calculated that if write-ups were eliminated from the rate bases of fifty-three of the state's utilities, the rates they charged would have provided a 15.3 percent return when 7 percent was the allowed return. The FTC made similar calculations for thirty-nine operating utilities for different years, resulting in sixty-three calculated returns. The average of the calculated sixty-three rates of return was 49.7 percent. Six of the returns exceeded 100 percent. These calculations, however, went beyond simply removing write-ups. Instead of the disparaged (but legal) use of replacement cost, the FTC valued the rate base using original cost. There was no explicit calculation of the effect of write-ups on rates, but there was no doubt they had caused increases. This view was shared by outside observers, individual regulators, and even industry executives.[44]

Although there was strong evidence that write-ups by operating companies led to higher rates, some argued that write-ups by holding companies also affected rates, even though the value of a holding company's assets were not part of a utility's rate base. There were, however, plausible mechanisms by which even those write-ups could have increased rates. When holding companies were in financial stress, they could and did respond by imposing part of that stress on operating companies, by pressing them to pay unjustified dividends, to make "upstream" loans to the holding company, or to purchase unneeded services or equipment. Most regulatory commissions had the power to restrain the last action, but they generally had no authority over either the dividends paid or loans made by a regulated operating company.[45] There were certainly instances when holding companies engaged in these practices but no clear evidence they led to higher rates.

Rate of Return

The FTC report did not provide mechanisms or examples that involved a holding company system manipulating a regulatory commission's allowed rate of return. Regulatory agencies did not have a great deal of latitude in setting the allowable rate of return, and holding companies apparently had little ability to manipulate that rate to the detriment of consumers.[46] Financial stress in an operating company caused by a holding company might have increased the rate it would have had to pay new bond or preferred stock holders. This could have led to an increase in the regulated rate of return, but the FTC did not consider this possibility.

The FTC report made clear that state regulation was structurally unable to prevent holding companies from using monopoly power to raise rates. Unlike the problems of investors, these were entirely due to a unique aspect of the industry's structure, the presence of regulated utilities. Solving this problem required federal action specific to public utility holding companies.

Public Relations and Political Influence

Opponents of holding companies had long held concerns about their manipulation of public opinion. The 1929–1935 FTC examination of this issue was the first, and a 480-page summary report was devoted to its findings.[47] The report concluded "that, measured by quantity, extent, and cost, this was probably the greatest peace-time propaganda campaign ever conducted by private interests in this country." Many techniques hid the role of the industry in the dissemination of information favorable to the industry, making it seem to come from disinterested experts. For the time, its sophistication and size was unparalleled. When it was revealed, many were shocked that such lengths and so many resources had been devoted to influencing public opinion, legislative outcomes, and even political elections. That response is evidence of how much attitudes have changed since 1935. Today, such behavior by large firms and industry groups is accepted as normal.

Different utility organizations at different levels were involved in the campaign. At the top were the National Electric Light Association, with the largest budget of any US industry trade group, and the American Gas Association, a smaller organization whose membership overlapped that of the NELA. The two trade organizations revived a previously moribund organization, the Joint Committee of National Utility Associations, which primarily was involved with lobbying but also contributed to the public relations/propaganda effort. State committees were set up to publicize the point of view of private utilities. They were supported entirely but clandestinely by privately owned utilities and given names such as "Committee

on Public Utility Information" or "Committee on Public Service Information." During World War I, Samuel Insull had become involved in creating
and disseminating war propaganda for the American cause, and in 1919,
he used this experience to create the first of these state committees in
Illinois.[48] Individual holding companies and operating utilities were also
actively involved in the efforts to increase public agreement with their
views, particularly among their own customers and investors.[49]

According to the NELA's director of publicity, every means of communication was used except "sky writing."[50] Many of the efforts were low-key.
Officials in operating utilities were encouraged to become active in civic
organizations and contribute financially to various civic, charitable, educational, and other organizations without any apparent strings attached.
Despite the absence of competition, utilities placed many advertisements
in newspapers ostensibly seeking to increase the use of electricity. Their
primary objective, however, was to gain goodwill and understanding from
the newspapers.[51] Utilities cultivated the support of bankers, especially in
small towns and rural areas, by making deposits that did not require payment of interest. Pacific Gas & Electric had such accounts with 230 different country banks.[52] Schoolchildren were encouraged to take tours of
utility plants. Teachers were invited to luncheons and dinners. Both teachers and students received summer jobs. Utility organizations provided
college professors with research grants and travel support, hired them as
consultants, and made financial contributions to their universities. College professors that had a "better understanding" of the utility position
were favored for support.

Residential utility customers were sold system securities in "customer
ownership" campaigns, where company employees or others went door-to-
door in residential areas using high-pressure sales techniques. The funds
thus obtained were beneficial, but the primary objective was to influence
public opinion. Customer ownership was characterized as the best form of
public ownership, unlike municipal ownership, which was called "political," or even "Bolshevik." Customers were encouraged to think of the utility as "their" company in which they had a say, although it was very rare for
their securities to have voting rights. Nevertheless, they frequently sided
with the utility on issues including rates. Congressional representatives
were informed of the large number of their constituents that owned utility
securities. Between 1920 and 1929, the value of securities sold customers
grew at an annual rate of almost 16 percent. Many of these residential users were unsophisticated investors who could not have been expected to

understand the risks involved.[53] After the economic crash, huge numbers of customer owners no longer sided with their utilities' positions.

Some of the efforts to influence public opinion were not so low-key. Variously named state committees, "news services," or "editorial services," with no apparent connection to utilities, would send newspapers stories and editorials that could be freely used and that supported the position of private utilities.[54] Sometimes privately owned utilities or their organizations targeted advertising support specifically to newspapers competing with those with "unfriendly" editorial positions.[55] Textbooks used in secondary schools and institutions of higher education received considerable attention and were evaluated based on their support for the viewpoints of private utilities. These evaluations then guided efforts to influence the choices of textbooks at all levels of education. Surveys of publishers and universities provided the names of authors preparing new textbooks who were contacted before their books were completed. Industry organizations designed new courses, offered them for use by schools and colleges, and used relationships with those schools and their faculty to help encourage their adoption.[56] The industry, primarily through the NELA and the Joint Committee, funded and widely distributed studies or other writings, including copyrighted books by seemingly disinterested experts, that supported the industry's position, without revealing the industry's role.[57]

Asserting the superiority of private ownership of utilities over ownership by any level of government was the primary objective of the industry's efforts.[58] Municipal utilities were characterized as inefficient, as charging high rates, and as being a drain on public resources. Surveys and analyses of individual municipal plants supporting these claims were publicized, but the FTC found that in at least some cases the information provided was incorrect, deceptive, or subject to different interpretations. They also found several cases of internal discussions among private utility organizations noting the necessity that these types of analyses exclude efficient and well-run municipally owned utilities. Publicizing the positive role holding companies played in the industry was also an objective of the campaign, as was opposition to federal regulation.[59] Exempt from criticism, however, was state regulation, depicted instead as highly effective at protecting consumers.[60]

Among advocates of government-owned utilities, one of the most popular was the Ontario Hydro-Electric Power Commission, known as Ontario Hydro. Initially created in 1906 to build and operate a transmission grid for power generated at Niagara Falls, in the 1920s it acquired generating

facilities from private companies and provided wholesale power to municipally owned utilities in the province. It grew into a large integrated network and the largest government-owned utility in North America. Until his
death in 1925, Ontario Hydro was led by Adam Beck, a charismatic and
able advocate of government ownership of utilities. Under his direction,
Ontario Hydro developed a reputation as an especially efficient and progressive utility. Its proximity to upper New York State and the differences
between its policies and prices and those of the neighboring US private
power companies put it at the center of the political controversy. Residential rates in Ontario were half or less those across the border, and Ontario
Hydro, unlike the private companies in New York, had a program of rural
electrification.[61]

The NELA and other private utility interests funded, often surrepti-
tiously, published policy studies by outside experts that disparaged Ontario Hydro. These studies' authors included W. S. Murray, the 1921 Superpower proposal author, and professors from Harvard and the University
of Minnesota. The most notorious example was published as a pamphlet
by the Smithsonian Institution that maintained that Ontario Hydro's rates
fell short of covering its costs by over $20 million. The author was Samuel
Wyer, depicted as an employee of the Smithsonian. The director of the
Smithsonian wrote a preface highlighting the findings that private ownership was superior to public ownership. Wyer was not a government employee. He had been paid by a private utility, and he had a history of
opposition to government power programs. Adam Beck asserted that the
report was filled with inaccuracies, and he issued a point-by-point refuta-
tion of its claims. In the furor that followed, the Smithsonian ceased making it available. The NELA, however, reprinted and distributed thousands
of copies.[62]

Even before the FTC had issued its final summary reports, the privately owned utilities and their holding companies took a number of de-
fensive moves in preparation for the response they knew these finding
would provoke. In 1933, the NELA was dissolved and replaced by the Edison Electric Institute (EEI). The new organization moved its headquarters to New York, forswore any lobbying or public relations activity, and
began operations with a budget that was only 46 percent of the NELA's
previous annual budget. The new organization pledged to confine itself to
the collection and dissemination of statistical data, to promote research,
to serve as a forum for industry members, and to be a liaison to other
organizations. In addition, it would enforce disclosure standards on its

member companies.[63] In January of that same year, the Commonwealth & Southern holding company selected Wendell Willkie as its new president. Willkie was a product of neither the utility nor the financial industry. He had been a junior partner of a law firm that worked for the holding company, and his great political, diplomatic, and negotiating skills were exactly what a holding company needed in the new political environment. With Samuel Insull in disgrace and no longer capable of performing the task, Willkie quickly become a leading spokesperson for the industry.[64] These actions, however, proved little help to the industry in the storm that ensued.

Death Sentence

Even before he became president in 1933, power issues had been a major concern of Franklin D. Roosevelt. He brought to the governorship of New York a distrust of giant utility companies and the concentration of control they were acquiring in his state. He shared the feeling of many New Yorkers that state commission regulation was ineffective. In his inaugural address as governor on January 1, 1929, he talked about the issue of electricity and said, "I want to warn the people of this State against too hasty assumption that mere regulation by public service commissions is, in itself, a sure guarantee of protection of the interest of the consumer."[65] As governor, he ordered a study that showed the average monthly bill paid by New Yorkers was two to seven times that paid by the customers of Ontario Hydro.[66] In a speech on September 21, 1932, Roosevelt more forcefully spoke of the shortcomings of state regulatory commissions and advocated a number of holding company reforms that included federal regulation. He also made clear in that speech that he did not favor a government takeover of the industry and expected private companies to conduct most of the industry's future development. He did advocate, however, specific but limited roles for government-owned utilities. Such utilities would be a "birch rod" by offering electricity users the opportunity to replace their private utility if its service was unsatisfactory and a "yardstick" against which the service and rates of private utilities could be compared.[67]

In 1934, President Roosevelt created the National Power Policy Committee to formulate legislation on electric utilities. Robert Healy, who directed most of the FTC's holding company investigation, was one of its members. The committee had discussions with utility executives, bankers, academicians, business leaders, and others with expertise in public utility holding companies. It requested recommendations on legislation from all state commissioners but found no interest by them in federally legislated reform. A District of Columbia regulator recommended that the federal government simply act as a fact-finding agency for the state commissions.[68] Twenty-six states agreed to cooperate but offered no suggestions; twelve states did not even respond. Both the administration and most of Congress regarded some legislation as necessary, although there was no agreement on the shape it should take. The committee leaned toward reform legislation to address the problems with holding companies, but the president increasingly signaled his desire for a "death sentence" that would eliminate public utility holding companies.

In February 1935, essentially similar bills containing the death sentence, titled the Public Utility Act of 1935, were submitted to both houses of Congress. There were several parts to the bills. The first dealt with holding companies and gave the new Securities and Exchange Commission (SEC) broad regulatory powers over them. With few changes, that part of the bill eventually became the Public Utility Holding Company Act of 1935 (PUHCA). The second part of the bill, after significant changes, became the Federal Power Act of 1935. It expanded and revised the regulatory powers of the Federal Power Commission (FPC).

The death sentence mandated the dissolution of holding companies three years after the passage of the bill. Holding companies could avoid this fate by demonstrating that their existence was necessary for control of a multistate "single integrated public-utility system." They, however, would have to divest all operating companies not contiguous to that system. Holding companies whose operating companies were all in a single state avoided all provisions of the law. The waiting period was designed to enable holding companies to build such integrated systems, perhaps by swapping operating companies among each other, but very little of this occurred.[69]

Holding companies had played a needed role by forming large fully integrated networks. That role would end with their dissolution. The second portion of the bill addressed the need for continued development of such networks. The Federal Power Commission had authority only on issues involving hydropower, but the bill proposed a radical change in its role and its powers. The most consequential aspects of the bill directed the FPC to divide the nation into regions and design large multistate integrated networks to serve each region. Electric utilities would be required to operate as common carriers providing generation, exchange of electricity, and transmission services in response to any request. If possible, the FPC would enlist the voluntary cooperation of the separate utilities in the area in setting up the new networks. If necessary, the FPC could compel a utility to construct any facility needed by the network. The FPC was to become the nation's network builder.

The FPC solicitor's Senate testimony explained why the elimination of the holding companies created this need:

> I think you can illustrate a holding company, more or less, as a pasture fence that is built around a group of these utilities. . . . Now title I of the bill [the death sentence] proposes to tear down that pasture fence and to turn the stock loose, that stock being the operating companies. . . . Title II provides that as you tear down this fence you shall place in the Federal Power Commission supervision over the integration of these operating companies into strong regional systems.

His testimony noted that holding companies had been involved in the creation of larger integrated systems but characterized them as not fully effective in meeting the public interest.[70]

The reaction of the private utility industry to the bill was intense. Despite its weakened public standing, it mounted a campaign ahead of its time in the sophisticated use of media, including the new media, radio. The bill was emotionally depicted as imperiling both individuals and institutions by destroying the value of utility and holding company securities. The bill's proponents were depicted as socialists out to eliminate private enterprise throughout the economy. The industry's efforts with newspapers paid off; two-thirds of all newspaper editorials opposed the bill. Rumors were started that the president and members of the administration were suffering from health problems. Senators and members of Congress were flooded with telegrams and letters from constituents, many complaining that the bill threatened their life savings. The campaign was so intense it became the subject of a Senate investigation. The investigation showed that many of the letters and telegrams had been prepared by industry operatives and were signed with names taken randomly from telephone books. When a notorious holding-company president responded to a subpoena to testify by going into hiding, the industry's standing fell even further.[71]

The FTC's investigation showed that holding companies effectively undermined the very basis of state regulation. Perhaps surprisingly, state utility commissioners showed little interest in correcting this problem. As

noted earlier, they offered no ideas to the National Power Policy Committee on how state regulation could be made effective. Much of the first part
of the administration's bill, including the death sentence, was designed
to restore the effectiveness of state regulation. In their testimony, representatives from the state regulator's national association (NARUC), however, had little to say about the death sentence. State commissions were
depicted as having varied opinions on that issue, but the only telegram
from a commission included in the testimony indicated that it had not had
enough time to form an opinion. Some of NARUC's witnesses expressed
mild opposition to the death sentence, but overall, their position seemed
neutral: "We leave that to the wisdom of Congress."[72] Their only substan-
tive comments on the first part of the bill expressed their desire to clarify
that only holding companies, not operating companies, were to be subject
to federal regulation and that regulation would not include holding companies whose operating companies were in a single state.

State regulators, however, were energized by the perceived threat to
their power from the expansion of the FPC's authority contained in the
bill's second part. Prominent in NARUC's testimony was recollection of
the loss by state regulators over railroad rates, depicted as having brought
"wreck and ruin" to that industry. The new powers that were to be given the
FPC similarly "would result in the emasculation of the powers reserved to
the States."[73] This was not the first time NARUC had raised this concern. In
1928, when the FTC investigation was still under consideration, NARUC
expressed opposition to any investigation that might lead to "invasion by
the Federal Government" of their authority.[74] Unlike the first part of the
bill, the second part elicited many telegrams from individual state commissions united in their opposition to expansion of the FPC's authority.
State regulators conceded that the 1927 *Attleboro* decision necessitated
federal regulation of interstate wholesale rates, but that authority needed
to be kept to an absolute minimum. They even opposed having the FPC attempt to encourage voluntary integration, because "Electric energy is essentially a local commodity" and large networks would threaten the quality of local service. This despite the facts that in 1934, 17.5 percent of the
nation's electricity was transmitted between states, some states obtained
the majority of their electricity from out of state, and other states sent a
majority of the electricity they generated to other states. Nothing positive was conceded about large networks. The notion that electricity was
still just a local commodity was a pleasant dream for NARUC but a bad
dream for the nation.[75]

Designing and implementing an integrated regional network was unprecedented for any regulatory agency. Details about how this would have worked were never developed. State regulators were justified in believing that an FPC with the powers proposed by the administration's bill would threaten their authority. State regulators approved any new construction. Could this include construction ordered by the FPC? A network requires central control. How would that control have been exercised? Many decisions would have likely encountered opposition from the individual utilities affected. How would they have been handled? The impact of this proposal on the structure of the electric utility industry would have been more profound than any other aspect of the administration's bill. The resulting industry very possibly would have had private ownership without private control. Perhaps such an industry could have served the nation better than the one dominated by holding companies, but it also might have done much worse. It was a risky and radical policy proposal that never had a chance. It died in both legislative committees.

The death sentence did make it out of the committees, and the battle over it remained intense. Not until it passed both Houses of Congress on August 22 was it clear that it would become law. On August 26, President Roosevelt signed the Public Utilities Act of 1935, including the Public Utility Holding Company Act (PUHCA) and the much watered-down Federal Power Act.

The SEC was made responsible for enforcement of PUHCA. Any company that owned at least 10 percent of the voting stock of a utility or of a public utility holding company was presumed to be a public utility holding company. All nonexempt holding companies had to register with the SEC. A holding company was exempt if all its operating companies were in a single state or outside the United States. A holding company could gain exemption by reducing its holdings to a single state. Registered holding companies were made subject to a large set of new requirements. They had to provide periodic detailed financial reports to the SEC. They were required to receive prior approval from the SEC before issuing any new securities. Many of the practices condemned by the FTC were forbidden, including upstream loans from an operating company to its parent company, political contributions, customer-ownership campaigns, sales of services by a holding company to its subsidiaries, the use of no-par stock, and the use of classes of common stock with different voting rights. The SEC was to direct the structural and financial simplification of holding company systems. The death sentence was to become effective in 1938, but full implementation took decades because of court challenges and the complexity of interpreting some of the law's requirements. Those holding companies that remained controlling interstate networks were small remnants of previous systems. When an existing holding company had more than one interstate network, it was split into separate companies, each controlling a single network. Most holding companies simply dissolved. It was possible for new holding companies to form, if they were necessary for a new interstate integrated network, but the process was difficult and had to be done in full public view. Holding companies in other industries, including the telephone industry, were not affected.

The Federal Power Act of 1935 gave the FPC authority over interstate wholesale rates, extending its authority beyond hydropower, and changes were made to that authority. Little remained of its proposed role to create large integrated networks. In consultation with state utility commissions, the FPC was to create districts within which an integrated network would be beneficial and to encourage coordination and interconnection within a district. Under certain circumstances, the FPC could order the construction of transmission facilities, but only under conditions of war or emergency could it order their use, and that authority was temporary. Not only did the new laws fail to provide an effective mechanism for the integration of existing service areas into fully integrated networks, they may have had the opposite effect. Becoming classified as a public utility holding company subject to SEC regulation was sufficiently undesirable that it discouraged interconnections and may have even resulted in the severing of existing connections.[76]

Did PUHCA Go Too Far?

Reform of the electric utility industry, and the role played by holding companies, was clearly justified. Was it necessary or desirable, however, for holding companies to be eliminated as was done by the death sentence? Holding companies played an essential role meeting the industry's needs during its development. Without holding companies, the young industry would likely have had difficulty acquiring needed capital equipment when that need was large and rapidly growing. Outside investment was crucial, and holding companies eased access to those investors. Was that role still needed at the time the death sentence was passed? The industry's experience since 1935 suggests it was not. Holding companies, such as Electric

Bond & Share, also played an important role providing the early electric utilities access to managerial and technical expertise. Although holding companies accelerated access to those services, independent consulting firms were a superior alternative, one that holding companies eliminated. The primary beneficial role holding companies continued to offer the industry was their ability to form large fully integrated networks. The proposed expanded role for the FPC shows that some understood the importance of that role, but it was not a priority. The proposal was ill formed and had little political support. State regulators were quickly able to quash it using clearly defective arguments. The 1990s brought an equally radical restructuring to the industry discussed briefly in the conclusion. It is still not clear if that structure is an improvement over the one it replaced. In 1935, however, the only known workable method to create large integrated networks was to put operating utilities under common ownership, as was done by holding companies. The Superpower experience demonstrated the difficulty of creating such networks otherwise, regardless of the total benefits.

New federal legislation brought financial reform to many industries, eliminating many of the problems identified by FTC without requiring the elimination of holding companies. There were problems unique to public utility holding companies, and PUHCA addressed those problems with the SEC's regulation of registered holding companies. Holding companies could have continued to exist under those regulations, and some, such as the prohibition of upstream loans, the provision of services to operating utilities, and the use of no-par stock, would not have affected their ability to continue forming large integrated networks. Others, however, would have. Holding companies had a powerful way of gaining control over operating companies by quietly acquiring a controlling interest in a company's voting stock without getting the assent of, or even notifying, the company's existing management. State regulation protected each operating company from competition and financial losses. This gave operating companies a substantial negotiating advantage that would have made a negotiated acquisition more difficult and more expensive. Under PUHCA, any acquisition by a registered holding company required the prior approval of the SEC and the state regulatory commission. The hearings that would have preceded that approval would not only have given the management of operating companies advance warning, they also would have provided forums for any opposed to the acquisition.

Even without the death sentence, it would have likely taken decades before holding companies would have been able form the type of regional networks that were envisioned for the FPC. The transfer of huge numbers of operating companies would first have had to happen, with a dominant company in each region still to be determined. The three years provided by PUHCA was much too short for any substantial progress in this area to have occurred. Very little did occur, perhaps in part because of the energy the industry expended in fighting all of the other New Deal programs seen as threats (chapters 7 and 8). The swapping of operating companies would have required selling new securities, not an easy undertaking in the Great Depression or Second World War, particularly given the reputation and status of existing holding-company securities. Perhaps a method could have been devised to give the FPC expanded powers to accelerate regional consolidations. Once it had determined the boundaries of feasible regional networks, it and/or the SEC could have been given powers to facilitate or mandate the transfer of operating companies in each region to a single holding company. This would have allowed common ownership and control of those networks. These new federal powers would not have threatened state regulators as much as those proposed by the administration, but given their paranoia, state regulators would probably have vigorously opposed them.

PUHCA drastically changed the structure of the electric utility industry, but the change was reactionary, not innovative. It sought to return the industry's structure to that which existed at least fifteen years previously. It returned to state regulators the power that holding companies had taken from them. The FTC investigation included numerous criticisms of state legislation and state regulation, much not directly the result of holding companies, although their political influence may have been at least partially responsible. In some cases, the FTC accused states of having facilitated holding company abuses.[77] The committee hearings on the administration's proposals did not feature criticism of the states, and state regulators apparently exerted great influence over the outcome. Perhaps in response to the FTC's findings, a number of states adopted reforms that strengthened their systems of utility regulation.[78] Nevertheless, great barriers were erected to the creation of new multi-state fully integrated networks. Other New Deal programs brought changes to the industry's structure that were much more innovative. These are the subjects of the next two chapters.

CHAPTER SEVEN

Hydroelectricity and the Federal Government

Until the New Deal, the federal government's role in the electric utility industry was small and limited to issues involving hydropower, and its role in that area was controversial. Two large dams, Hoover Dam and Wilson Dam, brought considerable attention to that controversy. Many of the issues were the same as those that divided opinions about any role of the government in the utility industry. However, to many the unique characteristics of hydropower provided stronger justification for government involvement than in other parts of the industry. When the New Deal brought to power individuals suspicious of the role played by private ownership in the industry, hydropower became the justification for a substantial increase in the role played by the federal government.

The use of falling or moving water as an energy source was well known in ancient times.[1] Before electricity, the United States was using more waterpower than any other country. In 1880, grain mills, cotton mills, and the production of woolen goods depended primarily on this source of power.[2] Factories that wished to use waterpower were generally constrained to be near the locations where it was available. A turning point in the history of electric power and the electrical engineering profession occurred in 1895 when a power plant at Niagara Falls began operation. Its design and construction involved individuals and companies from all over the world. Its use of alternating current to transmit power twenty-two miles to the city of Buffalo contributed to the industry's switch from direct current. In 1904, hydro generators at Niagara produced 20 percent of the electricity used in the United States.[3]

The cost of constructing a hydroelectric plant was typically higher than that of a fuel plant with the same capacity, but once it was constructed, it needed no fuel. Not all hydroelectric plants could be economically justified, but a large plant could produce electricity at a very low total cost. Hydroelectric sites were often located at a distance from where electricity was needed, but if the price was low enough, industrial users would move to be near the site. Nevertheless, hydroelectricity usually required greater transmission capability and a larger network than steam-generated electricity.

The rate of the flow of water and the height of the fall ("head") at a hydroelectric site determines how much electricity can be produced. A dam can increase the head and its reservoir can smooth the availability of generating water. Nevertheless, seasonal and unpredictable fluctuations in precipitation affect the amount of electricity a hydroelectric site can produce at different times. Electricity is most valuable if its supply is dependable, but the amount of dependable electricity from a hydroelectric plant is determined by the amount it can produce when the flow of water is at a minimum. More electricity can be produced at other times, but it is only available periodically. Dependable, "primary," or "firm" power commands a much higher price than seasonal, "secondary," or "interruptible" power. Unusually high water levels may require allowing water through the dam, and the "dump" power it produces is sold at a very low price to those able to make opportunistic use of it. The value of hydroelectricity can be increased by converting secondary power to firm power even if total generation is unchanged. Steam plants can do this by making up for power lost when water levels are low. A fully integrated network interconnecting many hydro plants can provide benefits above those such a network provides to a steam-generation system (explained in the introduction). Fluctuations in the availability of power justify a high level of interconnection among utilities not in a fully integrated network. Regions that relied heavily on hydropower, such as the Southeast, had particularly robust transmission networks.[4]

The federal government became involved with dams even before electricity. The Constitution's Commerce Clause gave the federal government responsibility for navigable waterways and their tributaries that affected interstate or international shipping. Irrigation, which often required dams, came to be seen as a federal responsibility. Areas of states subject to floods could be protected by dams in other states, and providing these dams also came to be a federal responsibility. Finally, in the West, many dam sites were located on public lands under federal control. When electricity was seen as required for national defense, the federal government again became involved in its production.

In the early twentieth century, a conservation movement arose from concern that private interests would not manage natural resources in the public interest, especially those resources on public lands. Many of these conservationists, including President Teddy Roosevelt, were Progressive Republicans. Certainly not socialists, they nevertheless believed that federal involvement was necessary to ensure the public enjoyed the benefits of the nation's natural resources, including hydropower. There was concern, for example, that a privately owned utility that controlled a hydroelectric site would take all the benefits of the cheap electricity it produced rather than pass them on to electricity users. Theodore Roosevelt expressed the concern in a message to Congress:

> It is especially important that the development of waterpower should be guarded with the utmost care both by the National Government and by the States in order to protect the people against the upgrowth of monopoly and to insure to them a fair share in the benefits which will follow the development of this great asset which belongs to the people and should be controlled by them.[5]

Federal operation was a solution to this problem.

The view that the federal government should operate dams that generated power gained support in the 1920s from growing interest in multipurpose dams. Multipurpose dams could generate electricity but also provide other benefits, such as irrigation, flood control, or improved navigation. Operating such a dam required trade-offs between maximizing the value of the electricity produced and providing other benefits. This did not preclude the involvement of private interests in hydroelectric generation, but the details of that involvement had to be defined. When the federal government was operating a dam, it seemed a small step for it to generate electricity. Many who did not support government ownership of other electric utilities did see a federal role where hydropower was involved. On this issue, they joined those who favored a larger role for government-owned utilities in all aspects of the industry. In the late 1920s, at the same time the holding company issue became contentious, two huge multipurpose dam projects—Boulder Canyon Dam (officially named Hoover Dam in 1947) and Wilson Dam—provided major venues for the battle over the proper role of government and private industry in the electric utility industry.

Although there was a question about how hydroelectric sites should be managed, there was widespread agreement on the desirability of their development. In 1926, Congress directed the Army Corps of Engineers and the Federal Power Commission to provide cost estimates for studies "of those navigable streams of the United States, and their tributaries, whereon power development appears feasible and practicable."[6] The corps' response identified all navigable streams and provided estimates of the survey costs. This was provided in House Document 308. In 1927, Congress directed the corps to carry out the surveys, and the twenty-four subsequent reports became known as the "308 reports." They laid the ground for subsequent battles over the development of hydropower.[7]

Federal Regulation and the Federal Power Commission

The 1890 Rivers and Harbor Act may have been the first extension of federal control over navigable waterways. That act gave the secretary of war the power to remove structures that inhibited navigation.[8] The first hydroelectric facility was constructed in the same year; the demand for dams to produce electricity increased rapidly after that. An 1899 act required the prior approval of the secretary of war and of Congress before a private firm could construct a dam that might inhibit river navigation.[9] A series of laws enacted between 1891 and 1901 gave the secretary of the Interior authority to grant the use of public lands to private interests for water development, including hydroelectric facilities, but the term of such a grant was not specified. A grantee that undertook to build a hydroelectricity facility faced the risk of losing it before recouping those costs, discouraging such investment. In 1906, control of national forests was moved to the Agriculture Department, and the secretary of agriculture was given permitting authority over hydroelectric facilities in those forests. Three cabinet secretaries were now involved in giving approval for waterpower projects: agriculture, Interior, and war. The policies of the three departments often conflicted and tended to change with administrations. As the demand for new hydroelectric facilities continued to grow, Congress had to deal with an increasing number of bills requesting approval for specific projects. There was widespread agreement that the development of waterpower needed a more coherent policy and streamlined procedures, but there was continuous disagreement over exactly what terms should be imposed on those given permission to develop waterpower. Conservationists, progressives, and others who were suspicious of profit-seeking private power companies were pitted against private power interests and their sympathizers.[10]

In June 1920, Congress attempted to rectify these problems through passage of the Federal Power Act and its creation of a Federal Power Commission (FPC) made up of the secretaries of agriculture, war, and the Interior.[11] The new commission was given the power to issue licenses for the construction and operation of hydroelectric facilities on both navigable waterways and public land. These licenses were to be for a maximum of fifty years in length, at the end of which the commission could renew the license, assign it to someone else, or the government could take over the project upon payment of the value of the investment made by the licensee. The FPC also was given the duties and powers of many state utility commissions, including the power to set rates and regulate the issuance of securities for licensees in two situations: (1) there was no state commission, or (2) when the states involved could not come to an agreement on issues involving interstate power. It could prescribe accounting methods, hold hearings, and collect information under subpoena. It had the responsibility of determining the value of the investments made by licensees and their costs of operation. In cases where state regulation was absent, it had the power to expropriate excess profits. State and local government-owned utilities were given preference over private companies in the awarding of licenses. They also were exempt from certain fees private companies were required to pay.

In its first year of operation, the new commission received applications for projects whose total power capacity was over ten times that of all existing hydroelectric projects on public land or navigable waterways, evidence of the pent-up demand for hydroelectricity and the need for a streamlined licensing process. In the second year, new applications were a third higher than the first. The new commission, however, did not get the resources required to perform the functions it was assigned. The three cabinet secretaries already had significant responsibilities, and the law permitted the commission only a single employee, an executive secretary, although the president could also assign an officer from the Army Corps of Engineers. All other tasks had to be performed by the staffs of the three commissioners' departments. Congress took this approach of an ex-officio commission, rather than an independent commission such as the already-created Interstate Commerce Commission, because of hesitation to give the president so much authority over a power situation about which so much controversy existed.[12]

Flaws in the commission's setup quickly became apparent. Some of the skills required by the commission to do its work, such as the valuation of utility property, were not present in the staffs of the three departments. A department that loaned the FPC a staff member was still responsible for paying that individual from its own appropriations. The cabinet members gave priority to their primary duties and spent little time on the business of the FPC. The commission simply could not perform its duties. Although the fees it received were sufficient, before 1928 it was barred from using them to pay its staff. Even after 1928, it was unable to hire staff outside the three departments and could only reimburse those departments for their staff. Congress had been resistant to giving the FPC independent status, but the resulting dysfunction forced a change. At the urging of President Hoover, the Federal Waterpower Act of 1930 finally made the FPC an independent agency.[13] That act did not change the duties of the FPC, which remained confined to issues involving waterpower, but it completely reorganized its structure. With the advice and consent of the Senate, the president was given the power to appoint five commissioners for staggered terms, and the commission was given authority to hire its own staff. Passage of the Federal Power Act of 1935, discussed in chapter 6, extended the FPC's authority beyond hydropower to the regulation of all interstate wholesale power rates. Built on the foundation provided by hydropower, the roles of the federal government as both industry regulatory and industry participant were expanded.

Irrigation and the Arid West

Although fertile, much farmland west of the one hundredth meridian suffered from a chronic lack of water, and irrigation dramatically increased the value of that land. In the 1870s, a developer purchased land near Modesto, California, for sixty cents per acre. Water became available to the land a little over a decade later, and it sold for $86.66 per acre. Land near what is now Pasadena, California, sold for seven dollars an acre in the 1870s. Within a decade, irrigation made the land suitable for orchards, and it sold for $500 to $1,000 an acre.[14] Not surprisingly, a large number of different organizations became involved with creating irrigation works, including

private profit-seeking enterprises (sometimes subjected to state regulation), mutual associations, state government, and the Mormon Church.
These organizations were responsible for a significant amount of irrigation, but their efforts frequently ran into difficulties, including sectional
disputes. The federal government had long been active in improving rivers
and harbors in ways that mostly benefitted eastern states. As additional
western states joined the union, the political power of the West, particularly in the US Senate, increased, as did the argument that reclamation,
the term for irrigation, was a national, not a regional, issue. Westerners
wanted the federal government to make irrigation available in advance of
its need and thereby encourage settlement of the region.[15] With the support
of President Theodore Roosevelt, in 1902 Congress passed the Reclamation Act, also called the Newlands Act after the Nevada congressman who
was its guiding force. The sale of public lands created a reclamation fund to
be used by the secretary of the Interior to fund the construction of irrigation works.[16] A newly created Reclamation Service within the Department
of the Interior's United States Geological Service took responsibility for
administering the act. Although it remained in the Department of the Interior, in 1907 the Reclamation Service moved out of the Geological Service
and was renamed the Bureau of Reclamation.[17] Replenishment of the fund
was to come from those benefitting from the irrigation projects, who would
repay the construction costs over time, but without interest. The lack of
interest payments, defaults by farmers receiving water, and the increasing
cost of reclamation projects resulted in chronic underfunding, and, beginning in 1915, annual Congressional appropriations were needed to keep
it solvent.[18]

The 1902 act had no mention of hydropower, but electricity was needed
for construction and pumping, and some of the bureau's projects were quite
distant from existing towns or cities.[19] The hydroelectric generating facilities built to meet these needs sometimes produced more power than was
required, particularly after construction was complete. In 1906 Congress
passed the Town Sites Act, primarily concerned with provisions enabling
the construction of new towns on land within the boundaries of an irrigation project.[20] Section 5 of the law permitted the sale of surplus power
under ten-year agreements, although the law made clear that irrigation,
not power production, was the primary determinant of project operations.
The money received from the sale of power was to go into the Reclamation Fund's account for that particular project, reducing the amount to be

collected from farmers. The primacy given to irrigation over power production meant that the power generated from many projects would only
be available seasonally. This secondary power was generally sold to privately owned utilities at a very low price, but it could only play the role of a
supplemental power source. The bureau was not restricted in its authority
to build transmission facilities, but ideological and financial constraints inhibited their construction.[21]

The 1906 law stated that the Bureau of Reclamation should give preference in the sale of electricity for "municipal purposes." This was the
first iteration of the "municipal preference clause" that, in the 1920 Waterpower Act, gave preference to states and municipalities for licenses to
construct and operate dams. Later versions of the clause gave state and
local government agencies and utilities, along with nonprofit rural cooperatives, preferred access to electricity generated at federally operated
facilities.

A milestone in the changing role that electricity played in the Reclamation Bureau's activities occurred with the Salt River Project in Arizona,
initiated in 1903 and completed eight years later. The major construction
project was Roosevelt Dam, located in an isolated area thirty miles from
the nearest town and sixty miles from the nearest railway station, and neither of them were connected by roads to the construction site. The project
required two hundred thousand barrels of cement, and transportation costs
were high. The raw materials for cement were located near the site, and
constructing and using an on-site cement mill cut the cost of the cement
by more than fifty percent. The manufacture of cement required electricity. A diversionary canal initially provided water to a hydroelectric plant
for this purpose. In 1909, operation of hydroelectric generators within the
dam started, and in 1911, the dam was completed, then the world's highest
masonry dam impounding the world's largest artificial lake.[22] Because of
the size of the reservoir, the availabilies of both water and electricity were
not as subject to seasonal variation as at other projects, increasing its value.
The privately owned Pacific Gas and Electric utility purchased a significant
amount of the dam's surplus generation, and it became the primary source
of electricity used in Phoenix. The proceeds from the sale of the surplus
power provided a major subsidy to farmers for irrigation water. The revenue from Phoenix's electricity showed the importance urban, as well as
rural, populations could play in the bureau's projects. The growing urban
needs of Southern California played an important role in a hydroelectric
project far larger than that of Roosevelt Dam.

Boulder Canyon and Hoover Dam

For sheer beauty, few of the world's rivers can match the Colorado. Its primary tributaries are fed with melting snow and precipitation from Wyoming, Colorado, and New Mexico. From a mountain plateau, the pristine river's elevation drops rapidly through a series of spectacular canyons in Arizona and on the border between Arizona and Nevada, including the Grand Canyon. After leaving the canyons, the now meandering river marks the boundary between California and Arizona and Arizona and Mexico until it crosses the border into Mexico and empties into the Gulf of California. During late spring and early summer, melting snow resulted in a flood with a water flow sometimes one hundred times that which occurred during dry times. In extreme cases, the flow of water approximated that of Niagara Falls; floods with water flows half that rate were common.[23] Unpredictable flash floods occurred at other times. Enormous quantities of silt were carried the river's length and deposited in a large and fertile delta.[24] These silt deposits raised the riverbed, eventually resulting in the river overflowing and carving a new bed. In prehistoric times, the mouth of the gulf was north of its present location, but it was slowly filled by the deposited silt. The rising land created a natural dam that cut off the northern tip of the gulf, leaving a shallow saltwater lake, the Salton Sea, in a depression that lay below sea level. The sun slowly burned away the lake, leaving a brine-encrusted depression in the middle of the parched and desolate, but very fertile, Imperial Valley.[25]

In 1901, a privately built canal to the Imperial Valley that ran partly through Mexico's Alamo channel came into operation. It lasted only a few years before being destroyed by the river's deposits of silt, but the agricultural boom it caused proved the incredible agricultural productivity of the valley. In 1904, the company restored the canal with an inlet in Mexico, partly to avoid conflict with the Bureau of Reclamation. The riverbed was higher than the surrounding land, and cutting through it created an opportunity for disaster. That disaster occurred in 1905 when the torrential river tore through the canal's intake, completely changing the river's path. Instead of flowing to the gulf, the river plummeted down a waterfall that was moving towards the inlet and that eventually reached one hundred feet in height. From there the entire river poured unchecked into the Imperial Valley. The Southern Pacific Railroad had loaned money to the corporate owner of the dam and, after the disaster, acquired the company's stock. President Theodore Roosevelt exerted pressure on the railroad's president, E. H. Harriman, to solve the problem, and, after two years, an enormous effort involving thousands of workers returned the river to its previous course.

Most of the Imperial Valley was saved, but the devastation was extensive and the prehistoric Salton Sea was recreated. Problems with floods in the Imperial Valley recurred, and Congress authorized the construction of levees in Mexico to keep them in check. Silt deposits continued to raise the river's banks forcing ever-higher levees. The rising river elevation made the situation increasingly precarious, floods continued to occur, and maintaining control over the river's path required constant effort. The periodic excessive water flow was interspersed with water levels so low that the Imperial Valley was afflicted with drought. Because most of the system was in Mexico, the needs of farms there were given preference in water allocations during droughts. Funds from the US paid for the flood control measures and provided benefits to Mexican farmland, but Mexico was seen as impeding flood control. Frustrations with Mexico led to a demand for a new canal, an "all-American" canal, which would bring water from the Colorado to the Imperial Valley by a route entirely within the United States. Although efforts were made to construct such a canal, its completion remained a dream for over two decades.[26]

The burgeoning city of Los Angeles lay a few hundred miles to the west of the Colorado River. From 1890 to 1900, its population more than doubled. In the next decade, it more than tripled. The lack of water made this rate of growth unsustainable. In 1913, with a population of just over 300,000, the city built an aqueduct over 200 miles long to the Owens River on the eastern side of the Sierra Nevada Mountains. By the 1920s, the aqueduct's drainage of the Owens Valley angered the residents of that valley and led to protesters blowing up the aqueduct in what became known as the "water wars."[27] By then Los Angeles had already set its sights on the Colorado River and begun the planning for a 300-plus-mile aqueduct to transport its water to the city. The course of the river through the canyons provided excellent sites for the production of hydroelectricity, and the population growth in Southern California that the river's water would enable would provide a market for that power.

Irrigation, flood control, municipal water, and electricity were all justifications for damming the river. By 1921, the Federal Power Commission had received eleven applications to develop hydroelectric sites on the Colorado or its tributaries.[28] Congress directed the Bureau of Reclamation to study the problems of the Colorado River and irrigation of the Imperial Valley. The report was issued in 1922. The bureau recommended construction of a high dam in Boulder Canyon with the sale of electricity paying the dam's entire cost. An alternate site was identified in Black Canyon, but the Boulder site seemed preferable. The dam site was ultimately moved from Boulder to Black Canyon, but Boulder Canyon continued to provide the name for the project and the original name of the dam.[29] Also recommended was the construction of the All-American Canal from Laguna Dam, approximately fifteen miles from the Mexican border, to the Imperial Valley.[30]

The dam would affect seven states: Wyoming, Colorado, New Mexico, Utah, Nevada, Arizona, and California. An impediment to the project was the character of Western water rights. Under the doctrine of "prior appropriation," used by most Western states, if available water from a river was insufficient to meet all needs, those who had been using the river's water the longest had priority over those whose usage started later. With Los Angeles and the Imperial Valley, California would quickly be receiving a large share of the water from the river's development and would forever have priority over the other states in receiving that water. This made the others disinclined to support the project. The total benefit available to the seven states was large, but agreement required an interstate compact that would override prior appropriation and divide those benefits up in a way that met the concerns of the six states other than California. Any such compact required the involvement of the federal government and the approval of Congress. A law enabling an interstate compact was passed in 1921, and Secretary of Commerce Herbert Hoover led the negotiations as the representative of the federal government. Compared to the Superpower negotiations Secretary Hoover also led, these must have seemed relatively easy, and an agreed compact was signed by the seven states in November of the following year. That concord proved illusory. Before becoming effective, Congress and all seven state legislatures had to ratify the compact, which failed to fully deal with the central problem.[31]

The compact divided the territories of the seven states into two areas: an upper basin and a lower basin that consisted of most of Arizona, California, and Nevada. The compact gave each basin the perpetual right to use 7.5 million acre-feet of water each year. The compact, however, did not specify a division of the water within each basin. That would rely on prior appropriation or a future interstate compact.[32] The creation of the two basins met the concerns of the states in the upper basin that California would compromise their future, but it did not address the same concerns in the lower basin. Nevada had so little irrigable land that the issue of its share of the water was not difficult to handle. This was not the case, however, with Arizona, and the Arizona legislature refused to ratify the compact. Years of negotiations between California and Arizona over the division of the lower basin share never brought them close to agreement.[33] Congress solved the impasse in 1928 with passage of the Boulder Canyon Project Act, enabling the pact to go into effect despite Arizona's strenuous objections.[34]

The act made the role of electricity clear—its revenue was to pay for almost the entire construction and operation of the future Hoover Dam.[35] The generation from Hoover Dam was to be far greater than that from any previous federal project. Some maintained it would be thirty times that of all other federal projects combined.[36] Before funds could be appropriated for construction, signed contracts had to provide sufficient revenue covering all operating costs and repay the government for construction costs (plus interest) within fifty years. Any revenue above that amount was to be placed in a fund to pay for further development of the area. The storage and provision of water would provide some revenue, but there was no doubt that the revenue from electricity would be far greater. Under the initial contracts, the revenue for firm power alone was more than sufficient to meet the payment conditions.[37] Water and secondary power would provide extra revenue above the minimum requirement. The act left it to the secretary of the Interior to decide whether the government would operate the generating plant or lease its operation to others. Although contracts were to remain in force for fifty years, the price of electricity was to be adjusted periodically as justified by "competitive conditions." Those purchasing the electricity would have to construct their own transmission lines to the dam's switchboard.

The debate over private versus government ownership soon encompassed the proposed dam. Los Angeles owned the utility that distributed electricity to its residents, the Bureau of Power and Light. Prior to the completion of Hoover Dam, Los Angeles purchased wholesale power from the Southern California Edison Company, a privately owned utility serving most of southern California. By 1921, both Los Angeles and Southern California Edison had applied for permits to build a dam at Boulder Canyon.[38] The interstate compact proposal led the FPC to suspend processing Colorado River applications, and the Boulder Canyon Act withdrew the river from private development. Since the dam involved irrigation and flood control (which were given first priority) and the river crossed an international boundary, the justification for federal operation of the dam was strong. Generation by the federal government at Hoover Dam and sale directly to Los Angeles's municipal utility was exactly what the advocates of government ownership wanted. Such an arrangement would provide the opportunity to show that government ownership better served electricity users than private ownership. To the advocates of private ownership, this would amount to the federal government going into competition with private business, establishing a dangerous precedent.[39] No dam and no power were preferable to that. The public relations/propaganda arm of privately owned utilities used all of its techniques to oppose the project, including subsidizing and distributing books and pamphlets critical of the project and using its influence to turn the press to its side.[40] A congressman sympathetic to the concerns of the privately owned industry proposed construction of a smaller dam that would have been unable to produce much electricity. Herbert Hoover, a strong advocate of private utility ownership, initially agreed with this suggestion.[41] The attraction of using electricity to pay for the dam was irresistible, however. Passage of the act was a victory for the proponents of government power.

After the act's passage, the Department of the Interior requested proposals from those potentially interested in contracting for power, for a price initially set at 1.63 mills per kWh for firm power and 0.5 mills per kWh for secondary power. There were four major applicants and twenty-three others. Both the city of Los Angeles and Southern California Edison applied for all of the power produced at the dam. The Metropolitan Water District, a governmental unit formed to bring water to Los Angeles and other cities, needed electricity for pumping water. It proposed taking half the power. The state of Nevada wanted a third. Ultimately, the secretary of the Interior, after negotiations, divided the firm power. The act gave preference to the domestic use of water, and the Metropolitan Water District received the largest share of the electricity—36 percent as well as priority for any secondary power. Los Angeles's allocation was 13 percent, and the other municipalities collectively were allocated 6 percent. Southern California Edison and three smaller private utilities received 9 percent. Arizona and Nevada were given rights to 18 percent each, although neither had need for that much power, and complicated rules governed the allocation of the power they did not use.

The initial contracts, signed in April 1930 during the Hoover administration, reflected a desire to distance the federal government as much as possible from the appearance of generating electricity.[42] Rather than selling electricity, the federal government was to lease the use of falling water. The federal government was to provide the generating equipment, but Los Angeles and Southern California Edison were to operate it. That operation was not to be a joint effort; it was to be as if there were two independent power plants. Los Angeles was responsible for producing the power for all governmental units, including the other municipalities, the Metropolitan Water District, and the states of Arizona and Nevada. Southern California Edison was to generate all of the power used by private companies. Los Angeles and Southern California Edison had to pay for the cost of the generating equipment and for the cost of maintaining it, but those costs would be credited toward their purchases of power. All of the signatories were required to pay for the power they were allotted regardless of whether or not they actually took it. Three months later Congress appropriated the money to begin construction.

President Franklin Roosevelt dedicated Hoover Dam on September 30, 1935, but the power plant was not yet finished. It was not until June 1937 that the initial power contracts came into effect. By then President Roosevelt was in his second term, the nation had weathered the worst of the Great Depression, the Public Utility Holding Company Act had been passed, and the policies of the federal government toward the generation and sale of hydroelectricity had undergone an enormous change. At this point, Hoover Dam was a relic of a previous period—one in which electricity was seen primarily as a source of revenue available to subsidize other river development benefits. In 1928, the passage of the Boulder Canyon Project Act seemed to establish a new precedent for expanded federal government involvement as a participant in the electric power industry. By 1937, however, attitudes and policies had changed so much that Hoover Dam seemed a less important precedent.

Muscle Shoals and Wilson Dam

A few miles east of Knoxville, Tennessee, the Holston and French Broad Rivers combine to form the Tennessee River. Rain and snow that falls in 40,000 square miles in seven southern states flows through creeks and streams that empty into that river. Its 625-mile journey takes it through Chattanooga, where it loops south, crossing northern Alabama until it turns north back through Tennessee into Kentucky, where, near Paducah, its waters flow into the Ohio River shortly before that river joins the Mississippi. Its watershed includes areas with the highest rainfall east of the Mississippi, and hydropower was the source of over 80 percent of electricity generation in the seven states in 1932.[43] The traditional inhabitants of the mountainous eastern portion of the valley were "hillbillies," independent people who sided with the North in the Civil War. Although rich in natural beauty, that area was poor in almost every other respect. The western portion of the valley was flat cotton country. Slavery brought riches to some in antebellum days, but in the 1920s, most of those living in this area scratched out a meager living from the soil.

In its pristine state, the river's flow varied by a factor of one hundred, from extreme dry to extreme flood. Floods were common and often destructive, particularly in the area around Chattanooga. In a seventeen-mile stretch in northern Alabama, the river dropped about one hundred feet at Muscle Shoals. Low water exposed rocks in that area; high water created white-water rapids. That section of the river was unnavigable and cut off both Knoxville and Chattanooga from river traffic to the Mississippi. Efforts to solve the Muscle Shoals navigation problem through the construction of canals began in the 1820s and continued for seventy-five years with only partial success. In the late nineteenth century, the area's potential as a site for hydropower was recognized. In 1898, Congress first considered a bill granting authority for private development of the area. Many other proposals followed, as did federally financed investigations.[44] No real progress was made, however, until 1916 and the looming involvement of the United States in World War I.

Nitrates, compounds such as ammonia or sodium nitrate where nitrogen is "fixed" or combined with other elements, are an essential component of fertilizer and explosives, including gunpowder. Although nitrates are now universally synthesized by chemical processes that obtain nitrogen from the atmosphere, prior to World War I most of the nitrates used in the United States came from natural deposits available at only a single location, a region on the Pacific coast of South America. In 1883, Chile defeated Bolivia and Peru in the War of the Pacific and gained sole control of the territory containing these valuable deposits, and sodium nitrate became known as "Chile saltpeter." These deposits were depleting, and several countries were increasingly depending on synthetic nitrates. The primary method at the time for synthesizing nitrates was the cyanamid process, which required large amounts of electricity. There was no plant in the United States using this process, but a US company, American

Cyanamid, had such a plant in Canada using power from Niagara Falls. Although all of the output of this plant went to the United States, it was a tiny proportion of the country's total nitrate use. In 1913, the German chemist Fritz Haber developed an alternative method for synthesizing ammonia that required much less electricity, and in 1918, he won the Nobel Prize for this work. Development of the techniques required to use the Haber process for the industrial production of nitrates occurred almost entirely in Germany. By the end of World War I, the Haber process was the most important source of nitrates in Germany, but no plant using the process existed outside that country.[45]

With US involvement in World War I increasingly likely, in 1916 Congress passed the National Defense Act.[46] A shortage of ships and concern that German submarines might interfere with access to Chile's nitrates prompted the inclusion in the act of a provision that empowered the president to investigate methods for synthesizing nitrates. The law also authorized constructing plants to synthesize nitrates and hydroelectric dams to provide those plants with the needed electricity. Interestingly, the act specifically directed that the government alone, without any private involvement, was to do all of this work. In April 1917, the United States entered the war. In September, Sheffield, Alabama, near Muscle Shoals, was chosen as the site for Nitrate Plant No. 2, producing nitrates using the cyanamid process. In December, the United States contracted for the construction of Nitrate Plant No. 1 to investigate the Haber process. In February 1918, construction of a dam at Muscle Shoals was authorized, as was construction of steam plants to provide power pending the completion of the dam. By the end of 1918, the war had ended. Nitrate Plant No. 2 was successfully tested too late to contribute to the war effort. Work at Nitrate Plant No. 1 was suspended in January 1919. Using funds already appropriated, work on the dam, named after President Wilson, continued until April 1921, but then stopped with the dam unfinished.[47]

What was to be done with the partially completed complex of projects constructed by the government at and near Muscle Shoals? In the 1920s, Congress may have devoted more time to fighting over the disposition of this white elephant than to any other single issue.[48] It was a battle that took many unexpected turns. At its heart was the ideological struggle over whether the provision of electric power was a private or government responsibility. The advocates of government ownership were led by Senator George W. Norris, the progressive Nebraska Republican who, as chair of the Senate Agricultural Committee, had considerable power over the issue in the Senate. Those who advocated the government turning over the facilities to private interests were dealt a blow by the Teapot Dome scandal that started in 1922. The federal government owned an oil field at Teapot Dome and leased the production rights to that field to a private company at terms very favorable to the company. Those terms were discovered to have been the result of enormous bribes paid to the secretary of the Interior, who was sentenced to prison for the crime.

The issues surrounding Wilson Dam, however, were more complicated than just the matter of whether the generation of electricity should be done by the government or by a private company. To some, the main issue was fertilizer, to others, electricity, and to others it was how the dam's benefits would be distributed among those in the Southeast. These issues created divisions within the groups favoring private or government operation that contributed to the inability of Congress to come to a decision. The paralysis caused by these disagreements imposed a large cost on the nation's economy by delaying for a decade the use of a valuable resource on which the federal government had already spent a considerable amount.

Initially the question of the disposition of the Muscle Shoals works involved fertilizer more than electricity. The debate shifted starting in 1921 when Henry Ford made a dramatic bid for the facilities. Ford's offer required the government at its expense to complete Wilson Dam and to complete another dam upstream from Wilson Dam.[49] Ford agreed to manufacture fertilizer at Plant No. 2 and to conduct research on other methods of producing fertilizer. He promised to sell fertilizer at a price that would limit his profit to 8 percent. Ford offered a very low payment for the dam and other facilities and, contrary to the Federal Waterpower Act, wanted a one-hundred-year lease on the dam, with preferential renewal rights. It became clear that Ford's intention was to build new factories that would use all of the power produced by the dam above that required for fertilizer. From the beginning, Ford maintained that altruism, patriotism, and the desire to provide farmers with cheap fertilizer were the sole motivations for his offer. In December 1921, Ford arrived at Muscle Shoals in his private railway car accompanied by Thomas Edison. Local citizens greeted him with enthusiasm. His stated plan was to build a new city on the banks of the Tennessee that would become one of the nation's industrial centers, larger than Detroit. The region would experience the greatest economic prosperity it had ever known. He would replace gold as the basis of finance with the "energy dollar." His proposal set off a speculative land bubble that attracted unscrupulous dealers and participants from across the nation.[50]

The battle over Ford's offer was intense. It generally had the support of farmers who were anxious for cheaper fertilizer and who accepted the argument that the Ford offer was opposed by a "fertilizer trust" responsible for high prices. However, Ford's commitment to fertilizer eventually became questionable. Ford's attack on Wall Street and his criticism of the "power trust" appealed to many progressives. Private power companies, particularly Alabama Power, whose service area included the region around Muscle Shoals, opposed the Ford offer because it appeared to deny them access to the dam's cheap electricity. Southern Democrats generally supported Ford's offer because of the beneficial effect the promised new industry would have on the region's economy. Some in the Harding and Coolidge administrations supported the Ford proposal; others opposed it, believing it to be a poor financial bargain for the government. Some manufacturers in the area opposed the proposal because of concern that Ford's plan denied them access to the dam's power. Some manufacturers outside the area opposed the proposal as being essentially an unfair government subsidy to a potential competitor. On two occasions, the House voted to accept Ford's offer, but Norris was successful in preventing the Senate from confirming that vote. In October 1924, Ford suddenly withdrew his offer, criticizing Congress for its indecision and for having played into the hands of Wall Street and the power trust.[51]

During the period when Ford's offer had been before the Congress, Norris had consistently advocated that a government-owned corporation operate the Muscle Shoals facility. With Ford's offer off the table, the primary issue became the question of whether the government (Norris's position) or private interests should operate the dam. This issue became comingled with the increasing attention paid to the issue of holding companies, which strengthened the position of those opposing turning over a public resource to private hands. Southern Democrats, who had been supporters of Ford's offer, split over whether public or private operation was preferred, but most remained anxious for a speedy resolution that would bring economic benefits to the region. Decision on the issue, however, remained elusive.

A remarkable series of votes that occurred during the two-week period between January 8, 1925, and January 15 of the same year demonstrated the literal inability of the Senate to come to a decision on Wilson Dam. Three proposals were before the Senate at that time: Norris's proposal for government operation, Oscar Underwood's (D–AL) to seek a private party to lease the facilities, and Wesley Jones's (R–WA) proposal to send the issue to a commission to study the issue and report later to the Congress. Initially the Senate approved the Underwood proposal for private operation. A few days later an amendment to the bill passed, sending the issue to a study commission, Jones's position. Next, an amendment for government operation, Norris's position, was approved. An amendment for private operation then passed, bringing the Senate back to its first position. That cycle continued with the passage again of an amendment to send the issue to a study commission. It appeared that this cycle of each proposal being able to replace a previously approved proposal might continue forever. An apparent tactical error by Norris, however, resulted in a final decision to seek private operation. Surprisingly, this apparent fickleness on the part of the Senate was not the result of an inconsistency on the part of individual senators. Four groupings of senators had differences in the way they ordered the three proposals. Within each group, those preferences did not change over the cycle. Each time an amendment was offered, a combination of at least two of those groups would lead to its approval. The possibility of "cyclical voting" has long been known, but this is the only documented case of it actually occurring in a legislative body.[52]

Over the next four years a number of proposals to give control of the Muscle Shoals facilities to different private companies, including power companies and chemical companies, came before Congress. At times one or another of these proposals seemed close to passage, but objections to each company prevented acceptance. Norris continued offering bills for operation by a government-owned corporation. Some of his proposals concerned only Muscle Shoals, with power from the dam to be sold to private companies. Others were much more ambitious and involved the development of the entire Tennessee River system with projects whose objectives included flood control and navigation as well as fertilizer manufacture and electricity generation. At times, the prospects for one of his proposals appeared bright, but Congress remained unable to make a decision. In September 1928 and again in February 1931, comparatively modest proposals from Norris actually passed both houses of Congress. The first received a pocket veto from President Coolidge. After the second was sent to President Hoover, Norris offered to resign from the Senate if Hoover would sign the bill. Hoover vetoed it with these words:

> I am firmly opposed to the Government entering into any business the major purpose of which is competition with our citizens. . . . The power problem is not to be solved by the Federal Government going into the power business. . . .

This bill would launch the Federal government . . . upon a basis of competition instead of by the proper Government function of regulation for the protection of all the people. I hesitate to contemplate the future of our institutions, of our Government, and of our country if the preoccupation of its officials is to be no longer the promotion of justice and equal opportunity but is to be devoted to barter in the markets. That is not liberalism, it is degeneration.[53]

Hoover's position and the deadlock that had gripped Congress for a decade was broken by the 1932 landslide election of Franklin Roosevelt. In January 1933, Roosevelt and Norris paid a joint visit to Wilson Dam. In May of that year, during the new administration's first one-hundred days, Congress passed the Tennessee Valley Authority Act, creating a government agency whose mission included and exceeded all that Norris had ever wanted.[54]

The Tennessee Valley Authority

The creation of the Tennessee Valley Authority (TVA) ended the congressional deadlock over Wilson Dam, but it increased the intensity of the struggle over government versus private operation of electric utilities. The TVA became the focus and the climax of that struggle. Its role as a "yardstick" and "birch rod" inflamed privately owned utilities that saw it as an unfair existential threat. Despite the damage done to their power and influence by the ongoing holding company issue, their struggle against the TVA significantly delayed its ability to fully implement an electric power program and made it a center of emotional controversy for decades. Even before its program was fully implemented, private companies maintained that the existence of the TVA damaged them.[55] The TVA's power policy, issued August 25, 1933, stated that the TVA directors should give "serious consideration" to the harm its program might do to private utilities, but the determining factor would be the public interest in having low-cost power.[56] During Roosevelt's presidency, the private utilities ultimately lost nearly all of their battles with the TVA. Their position resurged after World War II but largely ended after a compromise that insulated the TVA from the power of Congress and protected private utilities from any further TVA threat.[57] More than a half century of ideological battling over the proper ownership of electric utilities began then to recede from the nation's political consciousness.

At the time the TVA Act was considered, the private industry had been battered, but it still was able to mount an opposition. Perhaps its most effective leader was Wendell Willkie, the former corporate lawyer who had become president of Commonwealth & Southern (C&S), a subholding company of the United Corporation. Willkie was amiable and articulate, a progressive reformer within the industry, who would not defend the bad behavior that the FTC had revealed. One of the operating companies controlled by C&S was the Tennessee Electric Power Co., whose large service area encompassed a considerable part of Tennessee. Other operating companies also controlled by C&S, whose service areas were partly in the Tennessee River Basin, were Alabama Power, Mississippi Power, and Georgia Power, all states where the TVA first began providing electricity. Originally a Democrat who had voted for Roosevelt, Willkie became Roosevelt's Republican opponent in the 1940 presidential election. Willkie's political skill enabled him to identify those battles that were core to his company's concerns and those that were not. His positions appeared reasonable by compromising on emotionally polarizing issues. In one statement he said, "The utilities have no God-given charter for existing. . . . The only question which the people are interested in . . . is: What is the best way to generate and distribute electric power economically?"[58] He knew when to stop fighting battles that already had been lost. Ceding the issue of government generation, he even offered to let the TVA set the retail rates if his company were allowed to purchase and distribute the power, conditional on the TVA agreeing not to extend its own service into C&S territory.[59] Failing to achieve even that, he fought the continued growth of the TVA, arguing that federal subsidies were imposing costs on all Americans for the benefit of a single region and that the TVA set a dangerous precedent that could lead to government driving any private company in any industry out of business.[60]

In his message to Congress proposing the TVA, President Roosevelt described something new, "a corporation clothed with the power of government but possessed of the flexibility and initiative of private enterprise." Its goal "transcends mere power development. . . . It should be charged with the broadest duty of planning for the proper use, conservation, and development of the natural resources of the Tennessee River drainage basin and its adjoining territory . . ."[61] The Tennessee Valley Authority was unlike any other government agency created before or since. It was part of no executive department. Its headquarters were to be in the Tennessee Valley, not Washington. It was to be led by three full-time directors appointed by the president for staggered nine-year terms. A chair was to be appointed by the president but had no special power or authority over the other two. The directors could hire all other employees without regard to civil service regulations. The TVA could issue bonds to a specified limit backed by the full faith and credit of the United States. The new authority could exercise the power of eminent domain to acquire property. Challenges to TVA condemnation could be heard only in federal court. If such an appeal were made, rather than ratifying or overturning the price offered by the TVA, the law required the judge to make an independent determination of the value of the condemned property. Thus, a property owner who appealed the price offered by the TVA ran the risk of receiving an even lower price.[62]

The TVA was given the power to create a network that integrated not only electricity supply but also all of the area's natural resources. The act specified that the primary purpose of the dams built by the TVA was to be flood control and navigation, but it also authorized the generation of electricity. Electricity prices were to be as low as possible—not to cover all costs associated with a dam but only the portion attributable to power production. This reflected a huge change in Congress's attitude toward electricity pricing since the authorization of the Hoover and Salt River Dams, when revenue from the sale of electricity was to subsidize the dam's other functions.[63] In selling electricity, the authority was to give preference to government-owned utilities and agencies and nonprofit cooperatives over private power companies. Contracts for the sale of power to private companies had to contain a provision enabling them to be cancelled by the TVA at any time with five years' notice; this essentially guaranteed that a private company could not depend on TVA power for its own long-range planning. Residential and rural users of electricity were to have priority over industrial uses. The sale of electricity to industrial users was to be pursued only to enable lower prices for residential and rural users. The authority was to promote the increased use of electricity to better the lives of its users. The TVA's potential service area was undefined, bounded only by "transmission distance." At the time, this was about 350 miles. The area within 350 miles of Wilson Dam is huge and encompassed a large population, including several major and medium-sized cities (figure 7.1). As more dams were built, that area would enlarge. Clearly, the hydroelectric development of the Tennessee River system could not supply the total needs of this area, but a 1930 report by the chief engineer of the Army Corps of Engineers estimated that potential firm power production of such

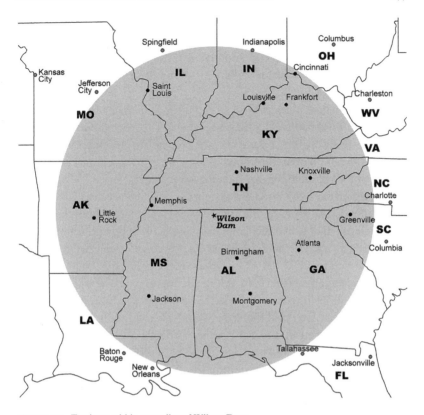

FIGURE 7.1. Territory within 350 miles of Wilson Dam

Source: Prepared by author

a development would be over four times the 1950 expected use of electricity within the Tennessee Valley.[64]

Regional planning was an idea that had acquired considerable interest and incorporated many worthwhile goals. Flood control, navigation, reforestation, and the development of plans "for the general welfare of the citizens . . . for the general purpose of fostering an orderly and proper physical, economic, and social development" were all specified by the act. The TVA Act devoted considerable attention to the production of fertilizer, and the TVA was given the responsibility of developing and encouraging techniques to improve farming.[65] The act insulated the board from political pressure to the extent that the TVA successfully resisted attempts to allow political patronage. In the short run, there was essentially no check on their decisions, and they received little guidance from the president. It was up to them to decide on the TVA's priorities and to design an administrative structure that could implement those decisions. Disagreement among the original board members in setting priorities led to a crisis in the young authority.[66]

Roosevelt chose Arthur Morgan to be the first chair of the TVA even before Congress had created it. Morgan was a respected hydraulic engineer who had designed a flood control project for the Miami River Valley in Ohio after a disastrous flood in 1913. Morgan was initially skeptical about the type of multipurpose dam that would characterize TVA projects. His project in Ohio consisted of a system of dams that remained dry except during floods. At each dam a granite block was placed with an inscription declaring that use of the dams "for power development or for storage would be a menace to the cities below," a statement often repeated by the TVA's detractors.[67]

Morgan was an unusual man. Although not religious, he was extremely demanding in his own moralistic behavior and in that of others. He had an engineer's attitude that there was a single best solution to every problem, and he carried that view to social problems as well as engineering problems. He was inclined to view those who disagreed with him as ignorant or self-centered and potentially as enemies. He was an elitist who believed that decisions should be made by enlightened individuals like himself acting in the best interests of others. Morgan believed that people should live in small communities and be motivated by a spirit of cooperation rather than competition. Above all, they should behave toward each other in a manner that was open and truthful. Morgan spent much of his life in a sequence of frustrating attempts to create utopias. Before coming to the TVA, he had been president of Antioch College. He rescued that school from certain demise and introduced the innovative system of work-study that became a model for other colleges. He ultimately became disenchanted with the school's students and faculty, and they with him, over what he saw as their failure to maintain the moral standards he expected.[68]

The first dam the TVA constructed was a little over twenty miles northwest of Knoxville at a tributary site that had received special attention from the Corps of Engineers in its 308 report for the Tennessee River. Its value came from the role its reservoir could play in controlling floods and increasing the system's proportion of firm power. The corps had developed an initial design for a dam that was completed by the Bureau of Reclamation.[69] The site had long been known as Cove Creek, but the TVA board named the new dam Norris Dam, honoring the senator who, more than any other individual, was responsible for the TVA. The standard practice in federal dams was for the construction to be done by contractors that competed in a bidding process. Morgan set the important precedent of having the work done at Norris Dam by the TVA's own employees, all of whom were expected to sign a code that dictated not only ethical behavior in their roles with the TVA but also his brand of morality in their personal life. He seemed to regard this as something that would self-evidently be accepted by all good people and was blind to its authoritarianism.[70] On typical federal projects, workers lived either in shantytowns under conditions that were dangerous and unhealthy or in company towns where living conditions were better, but still poor, and where they were required to pay exorbitant rents.[71] By contrast, Morgan called for a model planned community, also named Norris, which would provide homes that were modest but clean and safe with plumbing, electricity, and an electric stove. Winding footpaths connected the houses, and an all-electric school was provided. Workers were provided training programs, and bookmobiles brought them wholesome entertainment and education. Wages were above average for the region. It was another attempt to create a utopia.

The two other directors were Harcourt Morgan (no relation to Arthur) and David Lilienthal. Harcourt Morgan was an agricultural scientist. Born in Canada, he spent much of his career in Louisiana and Tennessee, where he rose to the position of president of the University of Tennessee. He brought a deep understanding of the Southeast and expertise in ways of improving the region's agriculture. Lilienthal, who grew up in Indiana, was a graduate of Harvard Law School, where he developed a passion for social justice and progressive causes. He started his career in labor law but became a master of public utility law. Prior to his appointment to the board of the TVA, he had served two years as the youngest-ever member of the Wisconsin Utilities Commission. He was a strong and effective advocate for government ownership of utilities, and leaders of privately owned utilities viewed him with enmity.

One of the roles expected of the TVA was that of a "yardstick." Many observers of the utility industry, both independent scholars and critics of private ownership of utilities, came to regard state regulation as ineffective in controlling utility rates. As a "yardstick," the TVA would charge rates against which those of privately owned utilities could be compared.[72] If found wanting, the example would compel private utilities to reduce their rates. It was an idea that appealed to Arthur Morgan, who advocated a method he thought would avoid rancor with private utilities. He wanted a good-will negotiation with Willkie to transfer a mutually agreed representative but limited portion of C&S's service area to the TVA for establishment of the yardstick. A scientific demonstration would then be conducted in which the TVA would operate under the same accounting and other ground rules as a private utility, enabling an objective comparison between it and the privately owned utility. Lilienthal, who harbored a deep mistrust of all Willkie represented, saw Morgan's position as threatening the TVA's very existence.[73]

Attempts to level somehow the playing field between the rates charged by existing government-owned and privately owned utilities had never been done. In some respects, the TVA did operate under rules more similar to those of a private company than was the case with most federal agencies. A 1935 amendment to the TVA Act required that it use the same Federal Power Commission's Uniform System of Accounts as privately owned utilities, that power projects be self-liquidating and self-supporting as soon as practicable, and that the rates charged for electricity produce gross revenue in excess of the total costs of producing power. As with a privately owned utility, the TVA was required to make audits of its operations public. The TVA was authorized to issue bonds to support the financing of the acquisition of distribution systems by municipalities and co-ops, but each instance required the prior approval of the Federal Power Commission.[74] Nevertheless, the TVA had numerous advantages not available to the region's privately owned utilities, and these compromised its ability to perform the yardstick function fairly. The cost to the TVA of raising money was much lower than that of privately owned utilities. The backing of its bonds by the full faith and credit of the federal government enabled it to borrow money at lower interest rates, and it did not have to pay dividends to stockholders. The use of multipurpose dams required that the TVA use as a cost of producing power only a portion of the dams' total costs. A private company that did not have a responsibility for navigation and flood control might have built a less expensive dam, but the revenues from the sale of electricity would have had to cover the dam's total costs. Even though the total cost of such a dam would have been lower, it likely would have been higher than the portion of the total cost of a multipurpose dam allocated to power production. In valuing facilities that it acquired from privately owned utilities, the TVA used replacement cost, the cost of then constructing a functionally equivalent (not identical) facility. The write-ups taken by many utilities justified a price lower than that on the private utilities' books, but the approximate 24 percent fall in prices between 1929 and 1933 made the TVA's valuations even lower. Essentially, the TVA took a write-down of the value of those assets, supporting lower rates than those of the privately owned utilities. The TVA and its distributors were not required to pay taxes. Both, however, made payments in lieu of taxes, but the amounts of those payments, particularly the ones made by the TVA, were always subject to controversy.[75] On the other hand, the impact of the TVA's special benefits on retail rates would have been through its wholesale rates. Privately owned power companies maintained that they could meet those rates, and in some cases, their wholesale rates were the same as the TVA's. Most of the retail price of electricity went to the distributors, not the TVA, and the extent of the advantages they had over the privately owned utilities is less clear. They may have had an advantage in being able to purchase an existing distribution system at an unfairly low price, but that is difficult to document. The TVA distributors did not have had to pay income taxes since they did not earn profits.[76]

Private corporations may have many goals, but serving the interests of shareholders is always important. The comparative vagueness of the TVA's goals led almost immediately to serious internal conflicts. The disagreement between Arthur Morgan and Lilienthal resulted in a public dispute in which Morgan came close to accusing his fellow directors of malfeasance. The near scandal this caused led to Morgan's firing by President Roosevelt. A subsequent congressional investigation (requested by Arthur Morgan) exonerated the other directors.[77]

The war among the original directors was not the only serious obstacle facing the young authority. It also faced a barrage of legal actions. By 1937, thirty-four suits had been filed against the TVA.[78] The Supreme Court ultimately decided two that challenged the constitutionality of the new organization. The first, filed in 1934, questioned the TVA's right to engage in the electric power business. A decision by a federal district court effectively shut down the TVA's power organization for several months until that decision was overturned on appeal, and the case went to the Supreme Court. Although the Supreme Court had already ruled against several New Deal programs, its decision, announced in February 1936, was in favor of the TVA.[79] That decision, however, was limited to power produced at Wilson Dam. A more fundamental challenge to the TVA's constitutionality was initiated in May 1936, when disagreement among the directors had become most intense, and proposals to come to some compromise with the privately owned utilities in the areas were under serious consideration by the Roosevelt administration. In that case, a number of privately owned power companies, led by C&S subsidiaries, argued that the TVA's power program went beyond a reasonable interpretation of "surplus power" and that its competition with private companies was illegal and in violation of the Fifth, Ninth, and Tenth Amendments to the Constitution. In December 1936, a federal district court issued a sweeping injunction that brought the TVA's expansion program to a halt and remained in effect for five months until vacated by an appeals court. The decision of the Supreme Court issued in January 1939 upheld the TVA's position that the Constitution did not protect privately owned utilities from competition.[80] A few days later, Commonwealth & Southern sold its entire Tennessee subsidiary, Tennessee Electric Power Company, to the TVA for $78 million. At this point, the TVA's future became secure and it had a service area sufficient to support the further development of its electricity program.[81]

If the use of electricity in an area increased, its average cost would fall. Lower prices for electricity would stimulate greater use, although not immediately. Increases in electricity usage largely depended on increased use of electricity-using appliances, including those for cooking, refrigeration, and heating. These appliances were not cheap and would not be quickly acquired and put into operation even after lower electricity prices made them more attractive. Whether the additional revenue and lower costs of increased use of electricity would offset the lost revenue from the lower price depended on how much usage increased under the lower price (demand elasticity). If a regulated utility lowered its prices, revenue and profits would fall in the short run. The increased use of electricity might (or might not) be sufficient to increase profits eventually, but vigilant regulation would take those away. By basing rates on a historical cost, regulation reduced the incentive for a utility to take the risk of experimenting with lower prices. The price decreases that did occur tended to follow increased electricity use rather than precede it. By contrast, the TVA was a new enterprise that was expected to operate initially at a loss. It was in a better position to take risks with rates, and it had a mandate to provide electricity at low cost and to increase its use.

The economic conditions caused by the Great Depression gave privately owned utilities an incentive to find some method of increasing electricity sales that would not have an adverse effect on revenue. Between 1929 and 1932, the current-dollar (nominal) price of electricity for all uses rose, although the nominal price for residential uses declined. Once adjusted for the period's deflation, however, the real price electricity for all uses was almost 37 percent higher, and for residential users it was over 11 percent higher. Over the same period, the use of electricity fell over 15 percent, and the number of customers fell almost 2 percent. Accompanying this was a shift back to isolated plants by some industrial users. Utilities were thus left with excess capacity, 30 to 40 percent in the region of the Tennessee Valley.[82] The extra cost of the additional electricity that unused capacity could provide was very low. Rate reductions were inevitable. In some areas, industrial users were successfully negotiating lower rates. Regulatory commissions in some states were pressing privately owned utilities to lower their rates, and four states gave their regulatory commissions emergency powers to order reduced rates unilaterally.[83] Since hydroelectricity does not have fuel costs, increasing the production of electricity to the capacity of hydroelectric generators was especially cheap. Engineers in a C&S subsidiary, Alabama Power, which depended almost wholly on hydropower, developed an ingenious solution that both protected the company's total revenue and offered a significant price inducement for customers to increase their use of electricity.

On September 2, 1933, shortly before the TVA announced its own retail rates, Alabama Power announced its new system of rates.[84] Two rate schedules were to be used, an "immediate," or standard, schedule and an "objective," or promotional, schedule with lower prices. Each month a customer's use of electricity was compared to the amount used in the same month of the previous year. A calculation was made of the customer's total bill for the current month using the lower objective rate and for the same month of the previous year using the higher immediate rate. If the calculated total of the current month's bill exceeded that of the previous year, the customer would pay the lower rates. If not, the higher rates would be used. To get the lower rates, a customer's use of electricity would have to increase enough that the total paid the utility under the lower rates would not have decreased. This rate scheme encouraged the increased use of electricity, but it put a limit on how far the utility's total revenue could fall. In its original formulation, the immediate (standard) rate would be reduced each year until, after three years, only the objective rate remained.[85] The rate system was made effective for Georgia Power on January 1, 1934. In a contract with the TVA, C&S's Tennessee subsidiary, Tennessee Electric Power, also agreed to adopt the rate structure, and it became effective the following month. Eventually, all C&S subsidiaries, and many other utilities throughout the country, adopted the same scheme. In addition to the promotional rate, even the immediate rate C&S introduced was lower than prior rates.[86]

Local distributors, not the TVA, were responsible for the retail sale of the TVA's electricity, and the TVA entered into a separate contract with each for the wholesale provision of electricity.[87] A number of features were common to all of these contracts, and the limitations they put on the financial operations of the distributors gave the TVA powers greater than those of state utility commissions. The contracts required that the TVA be the sole source of electricity for the distributor. In addition to specifying the wholesale rate paid to the TVA by the distributor, the contracts also specified the (relatively few) rate schedules the distributors could use. There were five standard rate schedules adopted by all distributors, although some distributors could impose surcharges, and cooperative distributors could charge higher rates to nonmembers. The contracts specified the accounting methods to be used by distributors. A municipality was required to segregate its utility's funds from all others. Revenue from the sale of electricity could be used only to cover the costs of the electrical distribution system; any excess had to be used to lower rates, which first happened in 1944. Payments to local governments in lieu of taxes were permitted at a level similar to that which a privately owned utility would have paid. The first community served by a TVA distributor was Tupelo, Mississippi, in February 1934, a little less than a year before the birth of the town's most famous native son, Elvis Presley.

As was the standard industry practice, the TVA's retail rates were based on kilowatt-hour usage and (for nonresidential users) maximum kilowatt use (demand).[88] The base residential rates, which applied to both rural and urban customers, received particular attention: 3¢ per kWh for the first fifty used in a month, 2¢ per kWh for the next 150, 1¢ for the next 200 kWh, 0.4¢ for the next 1,000 kWh, and 0.75¢ for all use beyond that. Some distributors were permitted by their contracts to apply surcharges. Two types of surcharges for residential users were common, although few, if any, municipal distributors were allowed to levy both. An "amortization" surcharge of one cent per kWh, with a minimum of twenty-five cents and a maximum of one dollar, enabled a distributor to repay the TVA for costs that the authority might have incurred to acquire the distribution system. Rural co-ops used a similar charge on customers to pay their membership fees, initially $100 ($1,750 in 2014 dollars). That fee could be paid in cash with a 20 percent discount.[89] These fees remained in place only long enough to cover the distributor's debt to the TVA or to cover fully an individual's membership fee. The TVA also allowed some distributors a uniform percentage "developmental" surcharge (generally 1 percent) if their revenue was initially insufficient to cover their costs, but this was rarely applied to residential customers. By 1940, the TVA had ninety-nine municipal distributors, twenty-four of which had used amortization charges. All of those, plus thirty-seven others, had developmental surcharges. Five of those had already discontinued the amortization charge by 1940, and one had discontinued the developmental surcharge. All of the thirty-nine rural distributors in existence by 1940 had both charges.[90]

In 1932, the national average price for residential electricity was 5.51¢ per kWh, and the average monthly residential use was fifty-one kWh. In the states within the Tennessee Valley, only residential users in Alabama and Mississippi paid a rate below the national average, 5.34¢, and consumers there had an above-average level of electricity use at sixty-two kWh per month.[91] Privately owned power companies and many contemporary observers felt that the new TVA prices were unrealistically low because there was no chance usage would rise to a level justifying such rates, but investors in privately owned utilities were worried. The day after the TVA's announcement, when the Dow Jones Industrial Average fell 1.9 percent, the price of a group of national utility stocks fell between 4.1 percent and 5.5 percent. The stock of Commonwealth & Southern fell by 4.3 percent. The value of utility bonds also fell.[92] The new rates were a major gamble by the directors of the TVA; to cover costs, average residential use would have to increase between 50 and 100 percent. If not, the TVA program would have appeared unrealistic and foolish.

Before selling power, the TVA had to come up with the rate schedules. Wilson Dam had already been paid for and was given to the TVA. The only costs created by using it to produce electricity were the relatively minor operating costs. Nevertheless, a cost for the dam had to be included in the TVA's rates for it to defend those rates as a yardstick against the inevitable criticism from privately owned utilities. Accurately determining that cost, however, was impossible. It required both assigning a total value to the dam and determining defensible portions to assign to power production versus navigation, flood control, and other uses.[93] Construction of the dam had begun more than a decade and a half earlier on a crash schedule to meet an anticipated wartime need. The dam lay half-finished for several years before finally being completed in 1925. The total amount spent on the dam was far more than would have been spent on a new dam. The

TVA estimated the cost of a new dam that met the same requirements and used that as a basis for its rates. Unsurprisingly, the figure it settled on received considerable criticism.[94] The next step was to determine a way of allocating the costs of that dam among its different uses. The TVA eventually conducted an allocation study, but that could not be completed in time to set the Tupelo rates. David Lilienthal, the director responsible for setting the rates, later claimed he deliberately overallocated the dam's value to electricity, using 50 percent when the allocation study set the correct amount as 40 percent.[95] In fact, the majority of the dam's value was a joint cost. Dividing it among its different uses was as meaningless an exercise as dividing the joint costs of an electric utility among different classes of users, as done by all utilities. In that case, as discussed in chapter 3, principles of price discrimination played a major role in those allocations. Whatever allocation the TVA made was sure to draw fire from its enemies. The only possible guide for that allocation was the vague notion of "reasonableness." In describing the process it used, the TVA explained the problem:

> A number of theories of cost allocation were studied carefully by the committee in its attempt to reach a conclusion as to the shares of the joint investment that should be assigned to the various functions. . . . the only definite portion that can be associated with any one purpose is the added cost made necessary by the inclusion of that purpose. . . . [allocating remaining costs] cannot always be determined by a common unit of measurement. The problem is one of judgment rather than scientific calculation.[96]

The advocates of private ownership quickly asserted, "TVA rates are politically made without reference to cost."[97]

Despite the rate reductions offered by the C&S subsidiaries, and the surcharges applied by some TVA distributors, the TVA rates were still significantly lower than were those of other utilities in the five-state area where TVA distributors operated before 1940. Table 7.1 compares the average prices charged by TVA distributors with a lower bound of those charged by non-TVA distributors and, separately, by C&S subsidiaries.[98] Depending on the level of consumption, residential rates charged by TVA distributors at the beginning of 1940 averaged between 18 percent and 36 percent below the rates charged by others in the same states. Nor was the discount restricted to residential rates. Commercial power rates charged by TVA distributors averaged between 22 percent and 43 percent below the rates charged by privately owned utilities in the same

TABLE 7.1 **Average 1940 TVA and non-TVA rates in Alabama, Georgia, Mississippi, and Tennessee**

Monthly usage	TVA	Non-TVA	Percentage reduction	Commonwealth & Southern subsidiaries
		Residential		
25 kWh	$0.83	$1.30	36.0%	$1.25
100 kWh	$2.78	$3.69	24.6%	$3.59
250 kWh	$5.38	$6.60	18.5%	$6.25
500 kWh	$7.37	$9.46	22.1%	$8.78
		Commercial power		
150 kWh	$4.95	$6.31	21.6%	$6.42
375 kWh	$10.40	$14.69	29.2%	$15.07
750 kWh	$17.26	$26.79	35.6%	$27.50
1,500 kWh	$27.51	$48.66	43.5%	$49.41
6,000 kWh	$91.57	$146.85	37.6%	$155.02

Source: US Federal Power Commission 1940, and see note 98.

states. Commonwealth & Southern's rates were not significantly different from those charged by the area's other privately owned utilities. Rates for large industrial users are not available.

The TVA had a mandate to increase the use of electricity, and its gamble with low rates gave it an incentive to see increased usage in both its area and in those areas it eventually expected to acquire. It encouraged adjacent privately owned utilities to lower their rates, and its early annual reports often praised those that did. Perhaps Tennessee Electric Power would have adopted the system of immediate and objective rates without being required to by its 1934 contract with the TVA. If the clause was not an inducement, it certainly was an endorsement by the TVA for privately owned utilities outside its territory to lower their rates. David Lilienthal fathered a new federal agency, the Electric Home and Farm Authority (EHFA), to encourage sales of large energy-using appliances, particularly to farmers. The EHFA negotiated with appliance manufacturers to lower prices and to develop new low-cost basic appliances. It also offered much more liberal installment-credit terms than the norm for the time and actively promoted the increased use of appliances. Although legally independent of the TVA, during its initial trial period it was run by the TVA directors. The EHFA program was made available outside the TVA's service area (in Tennessee, Georgia, Alabama, and Mississippi), but only if a contract was signed with the utility serving the area and the utility reduced its rates to an approved level. Willkie signed such an agreement, and the C&S subsidiaries Georgia Power and Tennessee Electric Power made the greatest use of the program. In congressional testimony, Arthur Morgan claimed that by the end of 1934, appliance sales in Alabama, Georgia, and Tennessee had quadrupled, with three-fourths of the increase outside the TVA's service area. In March 1935, the EHFA was transferred to the Reconstruction Finance Corporation and its program became available nationwide but had little impact on a national scale.[99]

It soon became clear that the TVA actually underestimated the growth in usage its service would provide. By June 1935, no community had been on TVA power for more than sixteen months. In eight of these early communities, five had already had average increases in electricity use of over 50 percent, and the community that had had TVA power the longest, Tupelo, increased average usage by over 164 percent. In addition, the number of residential customers in these communities increased between 6 and 44 percent. Distributors more than met the costs they bore. According to the TVA, by 1938 it had twenty-one municipal and nineteen cooperative distributors. Residential users served by them used an average of 121 kWh per month at an average price of 1.84¢.[100] For the nation as a whole in 1937, the corresponding figures were sixty-six kWh and 4.30¢.[101] By contrast, in Ottawa, where Ontario Hydro's rates were less than half those of the TVA, average residential use was 340 kWh.[102]

Since the TVA would only sell to distributors that either were owned by a municipality or were cooperatively owned, these distributors would have to displace the existing privately owned distributors, most of which were part of a large integrated system. This required a referendum where voters would decide if they preferred electricity from a privately owned or government-owned utility. For many it was a vote on whether or not they wanted a lower price for electricity. Despite the vigorous efforts of local private utilities to defeat these proposals, it is unsurprising that almost all the elections were in the TVA's favor.[103] A municipality that decided to switch service from a privately owned utility to a municipally owned utility served by the TVA either had to purchase the private utility's distribution system or construct a new parallel system that would render the older system worthless.

Although it made sense for the municipality to acquire the existing system and avoid the waste of duplicating an existing system, determining a price for the existing system was problematic. The extensive write-ups and inadequate depreciation taken by holding company systems in the 1920s resulted in substantially inflated book values of local distribution systems. However, because the distribution system of a municipality was typically part of a larger integrated utility system, the damage to the privately owned utility arising from its loss would exceed any accurate calculation of the value of the distribution system's physical assets.[104] Even worse for the privately owned utility, another federal agency, the Public Works Administration (PWA), handed municipalities a major bargaining advantage. If a municipality undertook the construction of a new distribution system, the PWA would cover 45 percent of the cost with an outright grant and would finance the remaining 55 percent with low-cost loans. PWA funds, however, were not available for the purchase by a municipality of an existing distribution system. Some municipalities obtained PWA awards before or during negotiations with the privately owned utility, enabling them to use the threat of constructing a new system as bargaining leverage.[105] However, many, if not most, of the private distribution systems were bought directly by the TVA, whose stated policy was to make every effort to avoid duplicating the facilities of a private power company, instead purchasing those facilities at an equitable price.[106] Private power companies proved adept at using the courts to prevent a municipal utility from going into operation until the TVA itself purchased the distribution system.[107]

The privately owned utilities made the reasonable argument that the effect of the TVA threat was to scare away investors. The Depression and the steep deflation that occurred during its first three years caused nominal interest rates to decline sharply. An inability to refinance bonds and preferred stock at the lower rates prevented the companies from enjoying a large decrease in their total costs.[108] The lack of statutory boundaries for the TVA contributed to investor concern with utilities in the large geographic region where they were threatened by the TVA. To Roosevelt, the TVA was only the beginning. In November 1934, Roosevelt said to a crowd in Tupelo, referring to the TVA, "what you are doing here is going to be copied in every State of Union before we get through."[109] Three years later, he proposed to Congress to divide the country into seven zones with a TVA-like authority in every zone. He included the idea in his tenth fireside chat with the nation.[110] All of this was going on during, or immediately after, consideration and passage of the Public Utilities Holding Company Act. The onslaught against private ownership of electric utilities seemed to be growing; private power companies had a solid basis for paranoia, and they fought back. Alabama Power used sometimes-untruthful advertisements in area newspapers to warn its customers of the dangers of service from the TVA, and Commonwealth & Southern argued their position in ads in areas they did not even serve.[111] In May 1936, Willkie wrote to Roosevelt, describing the situation as "practically open warfare."[112]

One interesting proposal emerged in 1936 in an attempt to achieve a truce between the TVA and the private companies, the power pool. The idea was to have a separate transmission authority that would purchase power from both the TVA's facilities and those owned by private utilities solely based on which plant could provide electricity at the lowest cost when it was needed. Those running the pool would then sell power at the same wholesale price to both government-owned and privately owned distributors. Britain had created a system similar to this in 1926 with the Central Electricity Board, which created and operated the British "national grid."[113] The proposal initially received positive comment from both the TVA and the private utilities, but the atmosphere of cooperation implementation of such a plan would have required was absent, and the proposal died within the year. Its proposed radical restructuring of the US electric power industry in the region echoed elements of the Giant Power proposal and anticipated the restructuring movement of the 1990s. Details of how the pool would have operated were never worked out.[114]

Was the TVA a Success?

The TVA has continued in existence to the present time, and the extent to which its operation has been viewed as successful has varied. During its first decade, it was unquestionably a public relations success. It benefitted from a very competent staff, it had unusually good labor relations, its dams were usually completed ahead of schedule, and architects praised their design. It won the hearts of the residents of the Tennessee Valley, and its press coverage was overwhelmingly positive. Even Wendell Willkie characterized it as an outstanding organization.[115] It offered substantially lower rates than neighboring privately owned utilities. TVA advocates argued that it caused those neighboring utilities to lower their rates as well, and there is strong evidence this claim was correct, at least in the case of residential rates. Between 1932 and 1937, the average price for residential electricity fell by almost 22 percent. All of the five states or groups of states that had the largest drops in residential rates charged by privately owned utilities were in or adjacent to the Tennessee Valley (table 7.2).

TABLE 7.2 **States with the greatest decline in residential rates, 1932–1937**

	Percentage change
Tennessee	*–46.5%*
South Carolina, Georgia, and Florida	*–35.6%*
Alabama and Mississippi	*–35.1%*
North Carolina	*–34.1%*
Kentucky	*–32.8%*
North Dakota	–32.3%
Wisconsin	–29.6%
Minnesota	–29.4%
Louisiana	–29.0%
United States	–21.9%

Sources: US Bureau of the Census 1934, 64–67; US Bureau of the Census 1939, 62–65
Notes: Data are for average revenue. Some data were provided only for groups of states. The total number of states or groups of states was 40.

There is, however, no evidence that those states differed from the others in the average rates charged for other uses of electricity.[116] Whether this was due to the TVA's role as a "birch rod" or "yardstick," or whether it was due to the TVA's direct efforts to persuade privately owned utilities to lower their rates, is difficult to say. The possibility of referenda in communities over whether to switch to TVA power created an incentive for privately owned utilities to reduce the perceived benefit of a switch for the class of users with the most votes. The TVA developed a productive and untapped resource, the Tennessee River, and created a large fully integrated network where none previously existed. It undoubtedly increased the efficiency of electricity supply for a large region of the country. The TVA had programs unrelated to its dams, including improving agricultural practices and land-use management. Although not considered here, they may also have been successful and valuable. Malaria was endemic to the region, the construction of dams and the lakes made it worse, and the TVA had an antimalaria program. Although it was not able completely to counteract the negative effects of new dams, it reduced them, it was regarded as innovative, and it was adopted by others.[117]

One of the TVA's mandates was to increase the use of electricity. It was successful at increasing the average use of electricity by residential users, but it is not clear how much effect it had on increasing the availability of electricity. Virtually all of the urban areas, including towns with only 250 people, were already served by privately owned utilities. Providing service to previously unserved rural areas was part of its mandate, but it is unclear how successful it was at achieving this. County-level data on farm electrification is available only from the decennial census of agriculture. In 1930, only 10 percent of US farms received service from an electric utility. For the seven states of the Tennessee Valley, the rate was even lower, 2.2 percent. The biggest change in the nation's rural electrification came with the establishment of the Rural Electrification Administration in 1935 (chapter 8). The rate of farm electrification for the nation as a whole had risen by over 23 percentage points by 1940. For the 44 counties in the seven states that had TVA service by 1937, the corresponding increase was 14.7 percentage points, while for the other counties it was 12.7 percentage points. This admittedly poor measure suggests that the TVA had a positive, but small, effect on the availability of electricity in rural areas in the Southeast.

Perhaps a more fundamental question concerns whether the TVA was successful in improving the economy of a chronically depressed region. On this question, two recent studies came to opposite conclusions.[118] Although the differences in their conclusions are not easily explained, there are strong reasons why this would be a difficult question to examine statistically. Ideally, such a study would compare the region's economic development with and without the TVA. Since we cannot observe the region without the TVA, the TVA region must be compared with another similar region employing statistical techniques to account for differences unrelated to the TVA. During the first part of the TVA's initial decade, the US economy was climbing out of the hole created by the Great Depression. Although the Depression continued for years, the rate of growth during that recovery was high. Then the nation entered World War II, and the economy grew even faster. During the first ten years of the TVA's existence, the economy of the nation as a whole (real per capita GDP) grew at an annualized rate of 8.8 percent. The next highest rate for any other decade in the nation's history was 3.7 percent.[119] Factors other than the TVA were likely more responsible than the TVA for the economic growth of its region and others. It would be surprising, however, for the large amounts of money spent by the TVA on dam construction and other projects to have had no effect on the depressed economy of the TVA's region, particularly since New Deal grants in general did stimulate local economies.[120]

The TVA had an extremely innovative organization for a government agency that gave it insulation from federal and state political pressure. It had a broad mandate that extended beyond electricity to implement new techniques in regional resource management. To its strongest advocates, the TVA was a blueprint for the even greater involvement of the federal government in the production of electricity as well as regional development and environmental conservation, with six more such authorities to follow. Considerable opposition came from privately owned electric utilities, but the TVA's independence from Washington also led to opposition within the Roosevelt administration.[121] Instead of being the first step of a new role for the federal government as producer of electricity, it was the apotheosis of that role. It is unlikely that the TVA could have ever been created outside of the first one hundred days of the New Deal. With the nation at the bottom of the Great Depression, it was far easier for a new president to get approval for a radically new economic program. The unprecedented low public standing of public utility holding companies blunted the ability of privately owned utilities to effectively oppose an enlarged role for government-owned utilities. By the eve of World War II, the TVA was firmly enough established that its continued existence was assured, but there would be no more. After the 1959 legislation that prevented the TVA from extending its territory, neither the TVA nor the TVA idea posed additional threats to privately owned utilities. Beyond its territory, the TVA had no lasting effect, and it settled down to being another firm operating within the nation's electric utility industry, albeit with a unique form of administrative structure and with responsibilities in other areas. It demonstrated that the distribution of electricity could be severed (at least somewhat) from its generation and transmission.

Was the TVA Necessary?

Could all that TVA achieved as an electric utility have been done by the existing privately owned utilities in the area? Certainly private utilities had the technical ability to unify development of an entire river system and create a fully integrated electricity supply network. However, there were significant barriers to overcome. Given the experience with other hydroelectric dams, construction and operation of all of the dams the TVA built on the Tennessee River would likely have been done by the federal government even with sale of the electricity generated to private utilities. This would have required greater involvement by Congress than was needed once the TVA was started. The multipurpose roles of the dams would have limited the ability of a private company to control generation from the individual dams. A single utility (or holding company) would have had to gain ownership over all the individual utilities serving the area. This would have been eased by the fact that most of the utilities serving the area were already owned by Commonwealth & Southern. That holding company had control of the multistate integrated network that had been built by Southeastern Power & Light adjacent to the TVA region. Conceivably, Tennessee Electric Power could have been incorporated into that network covering an area greater than the TVA's. Acquiring the other utilities would have been difficult, particularly given the barriers that were created by the Public Utility Holding Company Act. In the political atmosphere of the 1930s, private creation of a fully integrated electricity network in the Tennessee Valley would have been impossible. Even in the political climate of the 1920s, it would have been difficult. There is no way of knowing how the operation of a privately owned network would have compared to that of the TVA. Although it is easy to imagine a similar network created under private ownership, even given the many court delays, the TVA surely was able to do it more quickly.

The Columbia River and the Bonneville Power Administration

The Columbia River begins in the mountains of British Columbia and follows a twisted path that eventually leads it south to the city of Kennewick, Washington, where it is joined by the Snake River. There the combined river turns west, forming much of the border between the states of Washington and Oregon until it flows into the Pacific Ocean. The basin formed by the Snake and Columbia Rivers is a 259,000-square-mile web that covers a larger area than any country wholly in Europe and includes most of Idaho and Oregon, as well as Washington, a large part of Montana, and small parts of Wyoming, Utah, and Nevada. Large drops in elevation give the system a greater potential for hydroelectric generation than any other river system in the nation, but there was little exploitation of that tremendous potential until the New Deal. During the early years of Roosevelt's administration, construction began on two dams on the Columbia, Bonneville and Grand Coulee, whose generating capacities dwarfed any in the Tennessee Valley. The states with large territories in the Columbia River Basin already had the nation's lowest electricity rates and highest electricity use by residential consumers. In 1937, the average rates in Washington

State, Oregon, and Idaho were 2.5¢, 2.8¢, and 3.0¢ per kWh, respectively. The average for the nation as a whole was 4.3¢. Average electricity use by residential consumers in those states was between 1.7 and 1.8 times as much as that in the nation as a whole.

Like the Tennessee River system, the federal government eventually built and operated most of the dams on the Columbia River and sold the electricity produced by those dams at wholesale, with preferential access given to government agencies, government-owned utilities, and nonprofit cooperatives. In many ways the advantages given to these distributors over privately owned utilities was even stronger in the Columbia Valley than in the Tennessee Valley. Nevertheless, the organization of the government's role here was markedly different from that of the TVA and did not spark the level of private-utility opposition or the national controversy that characterized the beginnings of the TVA.

In 1932, the Corps of Engineers completed the 308 report for the Columbia River, describing a system of ten dams, including Bonneville, which was to be built at a site forty miles east of Portland, Oregon, on the lower portion of the river, and Grand Coulee about 450 miles upstream in Washington State.[122] In addition to power, dams below the mouth of the Snake River, such as Bonneville, were to enable improved navigation of the Columbia. Grand Coulee Dam would enable irrigation of the arid farmland east of the Cascade Mountains. Bonneville Dam's contribution to navigation led to its construction by the Corps of Engineers, while the benefits to irrigation by Grand Coulee Dam made the Bureau of Reclamation the logical agent for its construction. Differences in the interests of Washington and Oregon in the development of the Columbia River created some early tension. To these were added issues resulting from two rival federal dam-building agencies located in different departments answering to different secretaries. In July of 1933, the PWA provided a grant of $63 million for Grand Coulee, an event that worried Oregonians who feared this might prevent construction of their dam. Two months later, however, the PWA awarded $20 million to get Bonneville started. Congress later authorized completion of both dams. Although work on Grand Coulee had started earlier, controversy over the type of dam to be built contributed delays to its construction. President Roosevelt dedicated Bonneville Dam on September 28, 1937, and it began producing power the following year. Grand Coulee was not finished until 1941.

When work on the two dams began, there was no decision about how to distribute the electricity they would produce. In 1935, a number of bills were introduced to create an administrative structure to perform this task, operate the dams, and continue to develop the river. There were major differences in the proposals on those issues as well as how the wholesale prices for electricity would be set.[123] One proposal would have created a Columbia Valley Authority very much along the lines of the TVA, with three directors appointed by the president and the authority to construct and operate dams, powerhouses, and transmission lines and sell the power produced. Opposition to a TVA-like independent directorship came from, among others, the secretary of the Interior. He advocated centralizing control over all federal hydroelectric projects in his department and preferred a single administrator answerable to him. Both the Corps of Engineers and the Bureau of Reclamation opposed losing control of the dams they were constructing and losing the prospect of constructing future dams. The TVA-like proposal received relatively little attention from Congress, and the controversial issues required another solution.

The pricing question divided those favoring higher prices for those farther from the dam against those favoring uniform prices over the area provided the hydroelectricity. The Corps of Engineers promoted distance-sensitive pricing and received support from business interests in Portland. One motivation for the corps' position was concern over the size of the available market for electricity. Basing prices on transmission distance would enable the lowest prices near the dam, which would encourage new industrial facilities to locate there. The short transmission distance from Bonneville Dam to Portland made distance-sensitive pricing attractive to residents of that city. It is hard to imagine, however, such a pricing system being used with an integrated hydroelectric development incorporating the entire river system. There was also controversy over who would market the electricity. The Corps of Engineers had no experience marketing power. Their practice was to have this done by a private utility, which would pay a fee for the privilege. The Bureau of Reclamation did have experience marketing power from its dams, but its practice was to use high electricity prices to subsidize other project functions, an approach no longer in favor. The dominant view was for neither of the agencies to be involved in marketing and for a separate agency to get that task.

The disagreement over the proper form of ownership for utilities was as present in the Pacific Northwest as in the rest of the country. Opponents of private ownerships had a significant presence in Oregon and even more in Washington State. The difficulties of farmers getting service from privately owned utilities was a driving force in both states, and the Grange was active in backing a new form of government-owned utility.[124] In 1930, Washington State authorized the creation of public utility districts (PUDs), and Oregon amended its constitution to authorize the similar people's utility districts. These operated as special-purpose governments established by referenda with elected leaders.[125] In many ways, they were the most powerful form of government-owned utility. In addition to powers to raise money through borrowing, they also had limited power to levy taxes, and they could acquire the facilities of privately owned utilities through eminent domain.[126] The PUDs were likely to be major beneficiaries of the hydroelectric development of the Columbia River, and their special powers made them a greater threat to privately owned utilities than were the municipal utilities or cooperatives in the Tennessee Valley. Remarkably, in none of the public hearings held by congressional committees on the different proposals for administering the electricity from Bonneville was testimony provided by an official representing the privately owned utility industry or an individual privately owned utility in the region. The primary witnesses opposed to the bills under consideration were generally local chambers of commerce, and their efforts consisted of attempting to delay passage by having hearings held not just in Washington, DC, but also in the Northwest.[127] Privately owned utilities did participate, however, in later hearings that considered amendments to the act.

C. Edward Magnusson, professor of electrical engineering and former head of that department at the University of Washington, offered one of the most interesting proposals during the discussions.[128] Reminiscent of the Giant Power proposal and the TVA power pool, he advocated dividing the electric utility industry into three separate parts to generate, transmit, and distribute electric power using the British Central Electricity Board as a model. He foresaw a national grid but suggested in the interim the formation of three regional grids, one of which would encompass the TVA area. In his proposal, the federal government would operate the transmission systems, which would connect all generating stations and distribution systems, regardless of ownership. The federal transmission system would purchase electricity from generators and sell it to distributors. The proposal he submitted to Congress was not fully developed, and there is no indication it received serious consideration.

In 1937, with the imminent completion of Bonneville Dam, Congress passed what was clearly intended to be a temporary solution. The Bonneville Power Act contained this unusual clause: "The form of administration herein established for the Bonneville project is intended to be provisional pending the establishment of a permanent administration for Bonneville and other projects in the Columbia River Basin." Nevertheless, the basic structure created by that act has remained in place. The new Bonneville Power Administration (BPA) was to be headed by an individual appointed by, and answerable to, the secretary of the Interior. The dam was to continue to be operated by the Corps of Engineers, but the administrator was given authority over all aspects of the operation of the generators and the determination of the need for additional generating equipment. The electricity generated by the corps was to be delivered to the BPA to be marketed. The BPA was given broad powers to construct transmission lines not just to deliver power to users but also to interconnect any generating facilities owned by the federal government or any other governmental unit regardless of location. The act granted it powers of eminent domain greater than those given the TVA; in addition to real estate, the BPA was given authority to use condemnation to take possession of privately owned transmission lines and substations. By executive order, President Roosevelt gave the BPA the authority to market the power produced at Grand Coulee Dam. Unlike the TVA, however, neither the act nor the executive order gave the BPA full authority to coordinate power production for dams on the river system or to decide how to operate the dams to meet their multiple objectives. The BPA, unlike the TVA, never gained the power to develop a fully integrated network in the Columbia River Valley.[129]

The BPA's independence was somewhat muted by the role given the Federal Power Commission, which in 1935 had been given regulatory authority over interstate wholesale power sales. In addition to requiring FPC approval for the wholesale rates set by the new power administration, the act also gave to the FPC the task of determining how to allocate the total costs of multipurpose dams between power production and other uses.[130] Vesting these powers in an agency separate from the BPA may have provided some protection to privately owned utilities that its rates would not be unfairly low, but it also protected the BPA from any such accusations.

The "preference" provisions of the act designed to benefit government-owned and cooperative distributors were, if anything, even stronger than those in the TVA Act. As in the TVA Act, contracts made by the BPA with privately owned utilities had to contain a clause enabling it to cancel those contracts with five-year notice if a government-owned or cooperative distributor needed that power. Furthermore, in order to protect the preferred distributors from even having to wait five years, the BPA was directed to withhold 50 percent of the energy produced at the Bonneville Dam until 1941 from any sales to privately owned utilities except on a temporary basis. If a government-owned or cooperative distributor that was still in the process of formation made a request for power, the BPA was required to give that distributor the time required to complete the legal process, including holding referenda and securing approval to issue bonds. Like the original TVA Act, the service area of the BPA was bounded only by "economic transmission distance." Nevertheless, the BPA was not as directly involved in the formation of municipally owned utilities or cooperatives as was the TVA. It did not try to take over entire areas previously served by privately owned utilities. Unlike the TVA, the BPA did not have the authority in the original act to acquire steam-generation plants, which would have made it easier to provide service to such areas. A number of proposed but rejected amendments would have given it that authority, backed by an expanded power of eminent domain not possessed by the TVA.[131]

There were many similarities in the initial policies of the BPA and those of the TVA. The contracts between the BPA and its first municipal distributors and PUDs were very similar to those used by the TVA. The 1939 contracts specified the retail rates to be used by the municipalities, and contracts with PUDs required them to agree to use a standard resale rate schedule then under development that eventually was the same as those used by municipalities. Those rates were lower than the TVA rates, and residential users of electricity paid less than commercial users.[132] The BPA also was given authority over the retail rates of privately owned utilities that purchased its power. In 1939, the BPA made a one-year agreement to sell power to Portland General Electric, a privately owned utility that served the Portland area. The contract was made for such a short time period because the BPA and the utility had not completed negotiations on the resale rates for the utility. The utility did, however, substantially reduce its rates after signing the contract.[133]

At least initially, the BPA appeared to offer a potential threat to privately owned utilities similar to or greater than that posed by the TVA. In particular, the use of eminent domain by both the BPA and the PUDs gave the proponents of government-owned utilities a tool unavailable in the Tennessee Valley. Although privately owned utilities were relatively quiet during the deliberations over the establishment of the BPA, they had long been active in fighting PUDs, especially in Oregon. The creation of the BPA did stimulate attempts to create the new districts. Between 1938 and 1940, there were forty-one referenda in Oregon to organize People's Utility Districts, but voters rejected two-thirds of them. Despite its proximity to Bonneville Dam, for example, over 70 percent of Portland's voters rejected a People's Utility District.[134] Residents of Washington State were more sympathetic to local government distribution of electricity, and PUDs were already firmly established. Efforts, however, to dislodge existing privately owned utilities met fierce opposition. Attempts by PUDs to purchase the distribution system of a privately owned utility were as rancorous as they had been in the early days of the TVA. Even the power of eminent domain proved less powerful than many had hoped. As had frequently happened in other parts of the country, privately owned utilities were able to effectively use the courts to defend their interests, including blocking or delaying condemnation proceedings.[135]

Giving preference to municipalities, PUDs, and cooperatives did not require their active promotion by the BPA. The first administrator, J. D. Ross, who previously led the municipally owned Seattle City Light, was a veteran of the region's struggles over ownership of public utilities and had emerged as the Pacific Northwest's leading advocate of government-owned and cooperative utilities. As BPA administrator, he was active in promoting growth in government and cooperative ownership of utilities. However, after less than a year and a half in office, he unexpectedly died, and his successors settled on a more neutral, passive approach to this issue.[136] Perhaps privately owned utilities were less threatened by the BPA than the TVA because the former had less effect on rates. In 1940, the preference customers of the BPA did have low residential rates, but the rates for other government-owned utilities were as low or lower, and the rates of some privately owned utilities were also lower than were those of some of the preference customers. The amount of power available from the Columbia River system was so great that even with the preference clause, it seemed likely that privately owned utilities would have access to a large proportion of the available power. Finally, although it had considerable power to do so, the BPA did not threaten to take service areas away from privately owned utilities, as had the TVA. In 1940, the BPA estimated that for fiscal year 1941 only 5.3 percent of the electricity it sold would go to government-owned utilities and cooperatives, while 39.3 percent would go to privately owned utilities. It expected industrial customers located near the Bonneville Dam to purchase most of the electricity, 54.5 percent.[137]

Hydroelectricity's Role after 1940

Hydroelectricity brought the federal government into the electric utility industry, and it was a major source of electricity generation before 1940. The federal government began construction on several dams before 1940 that were not completed until later, and World War II led to even more dams being constructed by the TVA, the Corps of Engineers, and the Bureau of Reclamation. Three federal Power Marketing Administrations were created to sell the power generated at these dams, including (in 2011) power generated at Hoover Dam. The organization of these agencies was modeled on the Bonneville Power Administration, which was grouped with them. They sold electricity only at wholesale, gave preference in the sale of power to governmental units and nonprofit cooperatives, and, with minor exceptions, were limited to marketing hydroelectricity. Hydroelectricity also resulted in the creation of state-owned utilities, including major producers of hydroelectricity in New York, Texas, and South Carolina. As in the case of the federal agencies, the state agencies also give preferential access to distributors owned by a local government or a nonprofit cooperative.

The continued impact of hydroelectricity on the importance of government-owned generation has declined with its importance as a source of electricity. The availability of hydroelectric sites is limited, and the best sites were taken early. As the use of electricity grew, other sources of generation increasingly dominated waterpower. Total generation from hydroelectric sites increased continuously until 1974, after which there was considerable year-to-year fluctuation without sustained increase. Before 1940, hydroelectricity usually provided one-third or more of all electricity generated in the United States. World War II brought new dams, and, in 1945, hydroelectricity accounted for 35.9 percent of US electricity generation. After that, hydroelectricity's share of total generation almost continuously declined. During the first decade of the twenty-first century, its share was only about 6 to 7 percent.[138]

Although the role of the federal government as a participant in the electric utility industry grew slightly after 1940 with the formation of the new Power Marketing Administrations, they brought no innovations to the structure of the industry. The declining importance of hydroelectricity prevented their formation from being a continuing threat to privately owned utilities, and there was little controversy over their establishment.

Before 1940, two proposals, the TVA power pool and Magnusson's proposal for the BPA, were offered that would have created a very new structure for electric utility industry by dividing generation, transmission, and distribution into three different classes of firms, but neither were accepted.

Hydroelectricity brought the federal government into the electric power industry as both participant and regulator. Its policies toward the electricity it generated provided a special benefit to local government and non-profit cooperative electricity distributors. The Tennessee Valley Authority brought a new type of organization to the electric utility industry. Although there were once efforts to duplicate that organization, its creation depended on the unique economic and political conditions immediately following Franklin Roosevelt's election to the presidency. There is some evidence that the TVA at least initially had a net positive effect, but this remains controversial, and any such benefits are likely to have been temporary. It is not at all clear whether the wider adoption of its organization would have been better or worse than the institutional structures that prevailed. It is an issue no longer of sufficient public or political interest to receive serious consideration. The TVA separated distribution from generation and transmission, as did the BPA. The BPA system also somewhat separated generation from transmission, each done by different federal agencies, although considerable control remained centered in the BPA.

Hydroelectricity was also the entry point for the federal government as industry regulator. This role for the federal government continued to grow after 1940. It was not hydroelectricity, however, that sustained that growth. Since 1940, federal regulation of the industry has increased because of such factors as environmental protection and nuclear safety, and new regulatory agencies have come to have significant authority over the electric utility industry.

Rural Electrification

B y the 1930s, electricity had profoundly changed the lives and work of 75 percent of Americans. Untouched, however, were the majority of the almost 25 percent of Americans who lived on farms. Figure 8.1 compares electrification rates in farm residences with those in urban areas and small towns. In 1930 only about 10 percent of farmers had electric utility service in their homes, compared to 85 percent of their city cousins. That was still a significant improvement over the situation in 1920, when fewer than 2 percent of farm dwellings received electric service. This low initial figure did enable the industry to claim considerable success in farm electrification. The number of farms with electric service in 1930 was 6.3 times the number in 1920, an annualized growth rate of 18.4 percent.[1] The continuous increase in the percentage of farms with electric service is evident from the graph, even during the worst years of the Great Depression. What is also evident in that chart is that the growth in farm electrification accelerated after 1935, the year that President Roosevelt created the Rural Electrification Administration (REA). Its impact on rural electrification revealed a serious flaw in the existing structure of the electric utility industry, and its creation eliminated that flaw.

Electricity could play a major role in farm operations, but its impact on the quality of farm life was dramatic. The lack of electricity not only deprived farm families of the benefits of electric lights over kerosene lanterns, it adversely affected sanitation, nutrition, and the toils of homemaking. Without electricity, it was difficult to have indoor plumbing. Electricity was not essential to have running water. Windmills or gasoline engines could pump water to elevated storage tanks for running water, but this was much more expensive and less convenient. In 1930, over 84 percent of farm households had to go outside to get water, and almost 92 percent used out-

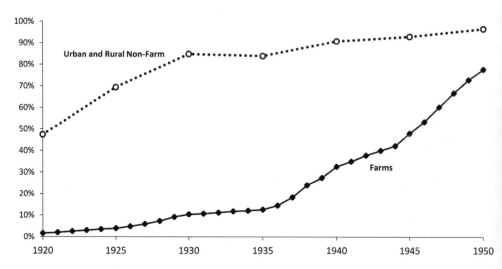

FIGURE 8.1. Percentage of dwelling units with electricity

Source: US Bureau of the Census 1975, part 2, 827

houses.[2] Outdoor toilets were unpleasant, and the diseases they brought, including typhoid, dysentery, and hookworm, imposed a toll on productivity.[3] Without refrigeration, the storage of food was more difficult, and without washing machines, irons, and vacuum cleaners (and indoor water) the work required to keep a home clean was far more time consuming. A farm homemaker in Virginia carried 150 pounds of water (about eighteen gallons) daily from a spring 271 feet from the house and sixty-five feet below it.[4] A 1934 study estimated that a family without indoor water had to devote over 300 hours a year to retrieving water and carrying it a cumulative distance exceeding 350 miles.[5] A Texas resident told the historian Robert Caro that every morning as a boy he would make seven trips to a well 300 feet from the house, each time with two four-gallon buckets, each weighing over thirty pounds, and his mother had to make more trips later.[6] Women were often depicted as bearing the greatest burden of not having electricity. To Caro, the lack of electricity on Texas farms meant "that the one almost universal characteristic of the women was that they were worn out before their time, that they were old beyond their years, old at forty, old at thirty-five, bent and stooped and tired."[7] Some blamed the migration of Americans from rural to urban areas on the lack of access to electricity in rural areas.[8] Given the stark differences between rural and urban life, it must have seemed surprising that anyone stayed on the farm.

There were alternatives to utility-supplied electricity for farms, but they had serious drawbacks. One obvious alternative was for a farmer to use an isolated plant. In the 1930 census, 9.1 percent of farms reported paying electric utility bills, but 14.3 percent reported using electric lighting. If the difference between the two percentages was due to isolated plants, there must have been over three hundred thousand farms using such plants. There were many manufacturers, and they advertised in farm journals. A 1923 advertisement from the company whose name became synonymous with isolated farm plants is shown in figure 8.2. The $335 price (excluding freight) is approximately $4,600 in 2014 dollars. This amounted to about 43 percent of average annual net income per farm in the United States in 1923.[9] Once the operating costs, including maintenance and depreciation, were included, the cost per kWh from such a plant was between 25.7¢ and 42.3¢ (about $3.50 to $5.80 in 2014 dollars).[10] These plants provided direct current, usually at low voltage, that was incompatible with many appliances. Some appliances would have required more power than they could supply and few appliances could have been run at the same time.

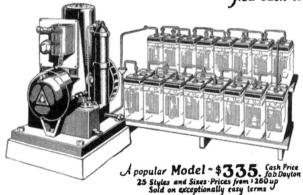

FIGURE 8.2. Advertisement for isolated plants for farms (digitally enhanced to improve clarity)

Source: *Farm Mechanics*, April 1923, 15

The Economic Problem of Farm Electrification

The cost to a utility of providing electricity to a farm was greater than
to an urban residence. Farm residences tended to be widely spaced, and
many were far from generating plants. $1,200 to as much as $5,000 ($16,000
to $84,000 in 2014 dollars) per mile were the often-quoted costs of dis-
tribution lines in rural areas.[11] Much of that cost was independent of the
amount of electricity distributed. Farms that did receive electricity from
a utility tended to be close to urban areas, with the minimum distance determined by the amount of electricity used on the farm. To avoid a very
high price, the farms receiving service had to purchase large amounts of
electricity. Figure 8.3 shows the percentage, by state, of farms with electric
service in 1935. California had the highest percentage at 53.9, and Mississippi had the lowest at 0.9 percent. States in the West and the Northeast
tended to have the highest rates of electrification for different reasons.
Many farms in the West required large amount of electricity for irrigation,
and those were states with low electricity prices. However, 40 percent of
farms in New Mexico and Wyoming were irrigated, but they tended to be

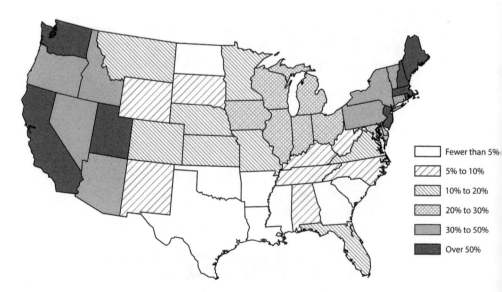

FIGURE 8.3. 1935 farm electrification rates by state

Source: Prepared by author from US Rural Electrification Administration 1940a, 54

so far from urban areas that electric utilities served fewer than ten percent of them. Farms in the Northeast tended to be close to urban areas. The relatively larger amounts of electricity used by poultry and dairy farms made them more likely than others to get electric service. For most farms, however, electricity's benefit to farm operations was less clear.

The cost of bringing the electricity to a farm was not the only cost of using it. Farm buildings had to be wired, and the appliances and machinery that would use the electricity had to be purchased. Once a farm got a utility connection, there would likely be a delay before the electricity was fully used. Once electricity is part of life, it becomes a necessity. For those who long lived without electricity, it was a luxury, and most farm incomes could not support many luxuries. Farm incomes tended to be both low and variable. Farmers did not participate in the economic boom of the 1920s, but the Depression hit them hard. In 1929, the economy as a whole was at a peak, but farmers did not participate in the prosperity. Net farm income had fallen by a third from its 1920 level, and it was less than 1 percent above that of the previous five years. During the Depression, average farm incomes fell to just over one-fourth their 1929 level, and they did not fully recover until 1942.[12] Many farmers had a strong desire for electric service, but the prospects that they would use much electricity were not bright.

The issue was emotional and personal for many farmers and contributed to the growing hostility many felt toward privately owned utilities. Stories abounded of farmers coming individually and in groups to utility executives begging for service, only to be flatly rejected, even when their need came from illness in the family and even when they were close to an existing power line.[13] The increasing integration of smaller utilities, often by holding companies, resulted in transmission lines that crossed rural areas where farmers without service could see them. Although this undoubtedly contributed to farmer frustration, serving farms from a high-voltage line would have required an expensive substation, and running a longer lower-voltage line from a city would have often been cheaper.

A number of foreign countries had much higher rates of farm electrification. In 1936, 80 percent of German farms were electrified. In both the Netherlands and France, over 98 percent of the entire population used electricity. Representatives of America's privately owned utilities were quick to point out crucial differences between the situation in the United States and in other countries.[14] Population density was lower in the United States than in most other countries, and farms were larger and more widely spaced.

In some countries, farmers lived in villages rather than on their farmland. Governments in a number of countries subsidized farm electrification. By 1936, France, Czechoslovakia, and Ontario had used direct government grants covering 50 percent or more of some of the construction costs of rural electrification. Sweden did not provide grants but did offer low-interest loans. The Netherlands had restrictions on the payments of dividends that resulted in utilities having reserve funds to support initial losses from extending service to rural areas.[15]

Early Industry Efforts to Electrify Farms

Despite the challenges of bringing electricity to farms, many within the electric utility industry were very aware of the plight of farmers and were hopeful of solving the problem of the economics of rural electrification. A 1930 report by Samuel Insull's holding company asserted, perhaps hyperbolically, "The advent of electrical energy into agriculture is undoubtedly an epochal human adventure, the ultimate consequences of which no man can even begin to estimate."[16] Optimists within the industry believed that electricity could become so important to farm operations that usage would be high enough to be self-supporting. Their efforts were devoted to identifying and encouraging electricity usage, particularly for electric motors.

The usefulness of power in a farming operation seems obvious, but the adoption of any mechanical power by American farmers occurred slowly. In 1926, International Harvester Company reported that animals provided 61 percent of farm power; tractors supplied 16 percent, and electric motors only 5.5 percent.[17] The 1930 census reported that only 13.5 percent of farms in the United States used tractors, and only 4.1 percent used electric motors for farm work. The total number of such motors amounted to one for every 16.3 farms.[18] This was not because electric equipment suitable for farm use did not exist. Many applications for electricity on the farm had been developed before the turn of the twentieth century and shortly thereafter.[19] These included even electric plows, using stationary motors, cables, and winches and were used by farms in Europe.

In 1910, the National Electric Light Association created the Committee on Electricity in Rural Districts. Its mission included identifying all uses for electricity in the nonurban areas then without access to utility-supplied electricity. In addition to farms, this included such potential customers as rock quarries, cement mills, and interurban railways. The committee's report at the NELA's next annual convention was sufficiently comprehensive to have included electric plows and "electric crop stimulation."[20] On the latter, "The subject is scarcely in a satisfactory state at this writing, but some experimenters have found good results with electric charges led directly into the soil, while others have obtained good results by arranging a network of electrical conductors at a height of three or four feet above the growing crop." The committee's recommendations did not include any actions by member utilities. Instead, it called for efforts by the association and by the federal government to educate farmers and efforts to get manufacturers of farm equipment to use designs enabling the use of electric motors.[21] Having established the usefulness of electricity in farm operations, the committee was also concerned with whether farms provided electricity would actually take full advantage of its benefits. In its 1912 report, the committee provided information about specific farm electrification programs carried out by its member companies and about the use of electricity on individual farms included in these programs.[22] In 1913, committees representing different sections of the country gave three separate reports. The committee for the eastern section advocated strongly for increased rural electrification, as did the committee for the western section, albeit with less passion. The committee for the nation's central section, however, concluded that "the 'farm business' as developed at present, does not furnish the central-station companies with returns commensurate with the necessary capital investment."[23]

For the next three years, the Committee on Progress reported on farm electrification at the NELA's annual conventions. All the reports dealt primarily with the beneficial use of electricity in farm operations, including, in 1914, the positive benefits of electrical stimulation of the soil.[24] Neither the use of electricity in the farm household nor the financial and technical issues of extending service to farms received attention. There was a sense of resignation about the topic. In the 1915 report, the committee noted that the benefits of electricity on the farm were receiving less attention. It blamed this on the fact that those benefits had been so well established that there was no longer novelty in them: "Why should the editor of an electrical journal waste space on the achieved when it can be devoted to the experimental and hypothetical?" Despite this, "it will be a long, long time before any one central station or power system has been [sic] able to live up to the almost infinite possibilities that lie in rural supply."[25] In 1916, "There is little that is new to add to the story of the electrification of

American farms, of which the number is in excess of 6,300,000, to many of which electric services and devices could be applied with advantage to all concerned."

The Costs of Supplying Electricity to Farms

In 1921, the NELA created the "Rural Lines Committee," which reported to the 1922 meeting and included separate reports from state committees.[26] Unlike the previous reports, this one dealt with the costs of serving rural areas and the rates that should be used in those areas. The committee opposed practices that might make the cost of rural service seem low. "In the discussion of the rural service problem, some have gone so far as to state that there are no overhead costs and that the bare cost of labor and material is the final answer. No more erroneous idea can be entertained. . . ." It rejected the position that development of rural electrification should entail losses even initially: "A large per cent of the rural customers now being served are receiving electric service at less than cost. This has established a bad precedent which will be difficult to overcome." Furthermore, rural customers should directly bear the utilities' initial investment in rural service by being assessed at least a portion of the cost of rural distribution lines before they were constructed. This was "about the only way utilities are able to build rural lines in the average rural territory at least at this time." The report provided five different methods for utilities to finance the cost of rural extensions. Three of them required advance payment by customers of all or part of the cost of rural extensions. Two had the utility paying the cost, but one of these required customers to purchase securities from the utility to finance that cost. The Ohio committee said, "we recommend that no company contribute any part of the cost of these lines," but the national committee was somewhat more accommodating on the issue. When customers paid for the lines, the committee advocated that ownership was to rest with the utilities, which would assume the costs of maintenance and replacement. The committee was dismissive of the idea of rural cooperative distributors: "It is quite generally conceded that the organization of a large number of small public utilities owned and operated by farmers is not desirable." The industry was undertaking a great deal of investment at that time in new generating facilities and in enlarging networks, of which both promised to lower costs and raise profits. The committee noted the greater importance of this investment over that required to provide service to farmers: "Financing of fundamental plant and line extensions must take precedence over the newer projects for the present."

The Wisconsin report included a detailed examination of the costs associated with seven rural extensions in Wisconsin for 1921. Two were built in 1917; the remaining five were built in 1920, the year before the report. The cost per mile of all utility equipment varied from $985 to $1,692 with an average of $1,226. Those served were required to pay advance amounts ranging from $254 to $452 depending on the number of customers served by a line. If additional electricity users later connected to the line, their payment for the line cost was divided up among those who had already paid. In addition to this initial cost, rural users paid the normal electricity charge paid by city users plus an extra rural charge. The extra rural charge was designed to cover depreciation on the rural line (whose construction cost had already been paid by the farmers), taxes, the additional losses in the distribution system above those in the city, and the extra maintenance, hazard, and operating costs associated with the rural line. For a rural user consuming the average amount of energy per month, the monthly bill was more than twice that charged in a city.

Although electricity use for farm operations had already been shown to make economic sense, the report advocated finding further applications that would be profitable to the farmer. Educating farmers so they would understand that using large amounts of electricity was in their own self-interest was extremely important. The government was seen as sharing a responsibility in these tasks. The report noted that government had been involved in both research into agricultural uses of electricity and in providing information to farmers about its benefits but argued that the utility industry, in concert with others, could make a real contribution.

There was considerable variation among privately owned utility executives in their views as to the economics of providing farmers with electricity, and individual views sometimes changed. However, the dominant view among privately owned utility executives for over a decade was that expressed by the report of the Rural Lines Committee, which viewed the issue of farm electrification as resting on these points:

1. A utility could not reasonably be expected to provide service to farmers unless the revenue from that service covered its cost, and there was good reason to believe it would not. Experience had shown that even after receiving service, most farmers did not use enough electricity.

2. Because of uncertainty over its return, investment in rural distribution lines was risky. For the investment to be made, much or all of that risk had to be assumed by the farmers themselves.

3. The cost of providing service to a farm included all (or nearly all) of the costs already allocated residential users in cities plus the additional higher costs associated with rural distribution.

Other reports added these:

4. There was ample evidence that the use of electricity in farm operations would increase productivity, lower costs, and, therefore, increase income. However, farmers who received electric service used it first (and often only) in their households. Although electrification would significantly improve the quality of life of farmers, this was a social goal, not an economic goal. It was not the responsibility of the utilities since "business is not a charitable institution."

5. Low income was the major reason for low electricity use by farmers.[27]

Despite these points being self-evident to most within the industry, they are based on dubious assumptions. The electric utility industry had long divided its users into classes, with a user's rates primarily determined by the class into which a user was placed. Although a utility's costs were allocated among these classes, and rates were set so that revenue from each class covered those costs, much of that allocation was essentially arbitrary, based largely on the sensitivity of electricity use to its price (chapter 3), often referred to as "value of service." The cost to a utility of providing service to users within the residential class varied in ways that were predictable but were not reflected in rates. At least in the short run, it would have been profitable to a utility to serve farms if the revenue received covered only the additional (marginal) costs of serving them.[28]

Basing the farmers' rates on residential rates seemed counter to the industry's position that operations, not household use, would be electricity's primary farm application. Farm operations had the characteristics of industrial, not residential use, and would have primarily occurred during the day in growing season. This was not when utilities then experienced peak demand, and the marginal costs of providing electricity using otherwise underutilized equipment would have been low.[29] Analyses from outside the privately owned sector of the industry maintained that the process of determining rural rates by starting with urban rates and adding the additional distribution costs was flawed. If, instead, rural rates were based on the total of all costs incurred in supplying rural customers, the rates would be lower.[30] By clinging to the view that usage levels determined prices (and not vice versa), the industry gave priority to requiring revenue from farmers immediately to cover costs rather than to developing a business providing service to farms. The average per-mile cost of constructing a rural distribution system would have been lower had a large system covering an entire area been built at once rather than piecemeal to individual farms or small groups of farms. This "building ahead of demand" received little attention. Some utilities lowered per-mile costs by using less expensive construction techniques in rural areas. Others dismissed that approach as threatening reliability, which was at least as important on the farm as in the city.[31]

It is not surprising that the first electricity used on farms went for household use rather than for farm operations. Farmers may have been told that using electricity in farm operations would lead to increased profits. Few, however, would have seen such use among their fellow farmers who, like them, had only just received utility service. Its immediate necessity would not have been pressing. By contrast, many well understood the benefits of using electricity to light the home, to do ironing, vacuuming, and washing, or to provide running water. The majority of Americans had long been able to take advantage of those benefits, and they were well represented in popular culture. Furthermore, some domestic appliances, notably electric ranges and water heaters, were major users of electricity even compared to farm implements. A significant reduction in the misery of farm wives could have had a direct long-run economic benefit. Even if electricity use just resulted in a larger cash cost of running the household, it would have provided an incentive for farmers to find ways of increasing their incomes, including by the use of electricity in farm operations. By improving health, household use of electricity might have improved farmers' productivity.

Researching the Prospects for Increasing Electricity Use on Farms

In 1923, the NELA, with the American Farm Bureau Federation, formed the Committee on the Relation of Electricity to Agriculture (CREA). The committee initially had twelve members, four from the NELA, three from the Farm Bureau, and one each from the American Society of Agriculture Engineers, a company that made isolated plants for farms, and the US

Departments of Agriculture, Commerce, and Interior. The membership increased over time; by 1932, there was no longer a representative from the producers of isolated plants, but instead of six, fourteen groups had positions on the national committee. Twenty-seven state committees were formed, generally consisting of representatives from the state's privately owned utilities, the Farm Bureau, and the local agricultural college. This is where the primary work occurred, with the national committee functioning as a clearinghouse and a publisher of the work of the state committees.[32] The major work consisted of studies undertaken by the state's agricultural college or an agricultural extension agency on the use of electricity in farm operations. In 1928, the Department of Agriculture indicated that CREA state committees were then running about ninety-five different projects.[33]

State CREAs also ran some experimental demonstration projects involving the electrification of actual farms. Farms that had not previously been supplied electricity were chosen and were provided with wiring and appliances to determine how much electricity would be used. The first such project was begun in Minnesota, soon after CREA's formation, in what was known as the "Red Wing" project, after a nearby dairying community. A line a little over six miles long served approximately twenty farms. Manufacturers of farm equipment loaned appliances and machinery to the test farms, each usage was separately metered, and the farmers kept records of use and time saved. Usage of electricity on those farms increased rapidly, and farm operating costs fell, as did the price of electricity. The results, however, did not transfer well to farming in general. There were only a small number of farms in the project, they were smaller in area than most farms, and, as dairy farms, both their potential to use electricity for farm operations and their money incomes were higher than most farms. The incomes of the Red Wing farmers rose, but this was during a period when the prices they received also rose. Their costs for electricity were quite high, over 6¢ per kWh in 1929, having fallen from over 8¢ in 1924 (about 82¢ and $1.10 in 2014 dollars). These results were not sufficiently positive to sell electric utilities on large-scale farm electrification.[34]

Alabama was the site of the largest electrification demonstration project under CREA auspices.[35] The president of Alabama Power, T. C. Martin, was genuinely aware of the benefits electrification would bring to Alabama farmers and wanted his company to help advance the region's development. The project was designed to provide a more representative test of the effects of bringing electrification to more types of farms than had the other demonstrations. Over 1,800 rural families were included in the test, but most were not farmers; the test included churches, lodges, and boarding schools. However, 379 farms were included, and an effort was made to reach all types of farms. No up-front connection charge was levied on the farmers, but unlike Red Wing and the other experiments, wiring, appliances, and farm equipment were not provided; farmers had to pay for those themselves. The project began with an agreement in 1924 between Alabama Power and Alabama Polytechnic Institute and ended in 1927. Alabama Power constructed several hundred miles of lines at a cost of $500,000 ($6.7 million in 2014 dollars). Only farmers able to purchase appliances could participate. County agents visited farms to provide technical help and advice on wiring, on the purchase and use of appliances and equipment, and on keeping records the project required. They would also accompany farmers to meetings with bankers to help secure financing. Community meetings were held, and literature was widely distributed.

Ninety-five percent of participants in the Alabama demonstration reported that electric lighting was the single benefit most valued. They also stressed the value of the relief from household drudgery that electricity brought, particularly carrying water and taking care of kerosene lanterns. Electricity was used for a variety of farm operations, particularly on dairy and poultry farms. The monthly use of electricity on those farms averaged 170 and 220 kWh, which exceeded the minimum of fifty kWh those in charge of the project thought was necessary for farm electrification to be self-supporting. Eighty percent of the farms in Alabama primarily grew cotton, however, and their average monthly consumption of only twenty kWh was disappointing. There was little electrical equipment then for use in cotton-farm operations, and those farms tended to grow only that crop. Although the cotton farmers chosen for the project were able to purchase appliances, most cotton farmers were too poor to afford them. The project's leaders concluded that the solution to the problem of farm electrification lay in fundamental changes in the structure of Alabama farming. This only reinforced the view within the industry that farm electrification could not be justified as a business proposition. Farmers in areas of Alabama with rural population densities typical for the nation paid between 3.9¢, which was lower than most utilities' rural rates, to 12¢ per kWh, which was high and would have discouraged use. Unlike the Red Wing experiment, they had to pay for wiring and appliance, and they were not targeted with as extensive an educational effort. Three years or less (depending on when the lines were completed) was not long enough to have seen a complete adjustment to electricity use.

An interesting study of the problem of rural electrification that came from outside the privately owned sector of the utility industry was included in the 1925 Pennsylvania report of the Giant Power Survey (also discussed in chapter 4).[36] One of the objectives of that project specified by the state legislature was the provision of "an abundant supply of electric current . . . for the . . . farms . . . of this Commonwealth," but the study found that less than 6 percent of the state's farms were provided electricity by a utility.[37] Of the 480 pages of the report, one-third was devoted to rural electrification. The study did not use experimental projects to determine electricity usage on farms previously without service. Instead, it collected data on the usage of electricity by farms that were already receiving that service. It also collected data on farm electrification in foreign countries and from utilities in other states in the United States (including the Red Wing project) and provided engineering studies of farm electricity use. Studies were also done on the costs of distributing electricity to farms and on techniques to reduce that cost. According to the study's director, Morris L. Cooke, the cost of providing service to seven-hundred fifty thousand unserved individuals, half of whom lived on farms, would have been less than $30 million. If spread over ten years, this would have amounted to 2 percent of the annual capital expenditures of Pennsylvania's utilities. A single private utility in Pennsylvania spent over $100 million in capital expenditures in 1923 alone.[38]

Upon completion of the report, the Giant Power Committee submitted several bills to the Pennsylvania legislature. Included among the proposed laws was a requirement that the Pennsylvania Public Service Commission establish uniform standard rates statewide, a provision for the cancellation of a utility's exclusive charter to serve rural territories where it was not doing so, and provisions enabling the creation of cooperatives and special-purpose governments to provide electricity (PUDs). These provisions were opposed by the privately owned utilities and were rejected by the legislature. The Public Service Commission did issue an order requiring utilities to provide service to a farmer who requested it. If three or more farmers per mile pledged to take the service, the utility had to bear the entire cost of constructing the new line. If fewer than three farmers made the pledge, the utility could charge the farmers for the line but had to cover the first $300 per farm of those costs. The order, however, was silent on the issue of rates.[39]

Unlike the studies performed by CREA, the authors of the Giant Power report did not see as their primary objective the determination of whether rural electrification could be profitable within a few years. Their assertion that farmers would use over 100 kWh per month would have justified providing them with service, but it is not clear that this level of usage would have been typical for American farmers or even for Pennsylvania farmers. The data gathered from different utilities showed the inverse relationship between the price of electricity and its use. Those data were interpreted as showing that that low rates were necessary to stimulate usage. Privately owned utilities, however, could interpret the same data on prices and usage as a reflection of the need for prices to be high when usage was low, reversing the direction of causality.

Some privately owned utilities were more active than most in providing electricity to rural areas. An interesting example of an unusual rural electrification policy of a privately owned utility was that of the Milwaukee Electric Railway and Light Company, whose low rates and simple rate schedules were commended by the Giant Power report. Of particular interest was the policy's recognition of the difficulty to a farmer of acquiring the financing needed to support electricity's initial use. The company did impose a line connection charge, although less than that of many other utilities. It estimated the total revenue prospective customers would provide over a three-year period and would cover all connection costs up to that amount. If a prospective customer or group of customers felt the company had underestimated their future use, the company would accept, in most cases, a guarantee that the customer would use at least the required amount of electricity. Alternatively, a customer could arrange to make the payment in installments for the following year at 6 percent interest. These policies eliminated delays in the extension of rural lines. If new customers made use of the same distribution lines within three years, for that three-year period, the original customers would receive half the bills paid by the new customers, encouraging customers to sign up their neighbors. The company offered to finance for one year (also at 6 percent interest) the cost to a farmer of wiring a house. In 1928, more than 20 percent of new customers took advantage of this offer. The utility would also sell appliances to a farmer on an installment plan. Farmers who received service from this utility continued to increase their use of electricity for at least seven years, by which time that usage had almost doubled.[40]

Why So Timid?

By 1935, privately owned utilities remained hesitant to take the risk of providing service to farmers. The great majority were unwilling to undertake the needed investments unless their costs were covered in advance.

The industry had undertaken considerable investigations into rural electrification, and the results of all of them confirmed that rural electrification would not quickly pay for itself. The industry seemed to recognize neither that adjustment to the availability of electricity took time nor that the conditions under which it was willing to provide service delayed that adjustment. The utilities were in a much better position to finance the needed investment, but they required that to be done by the farmers themselves. The policies of the Milwaukee Electric Railway and Light Company showed recognition of these issues, but even it was willing to provide only short-term financing where long-term financing was needed. Perhaps timidity could be understood had this been an industry reluctant to take risks in other areas, but the opposite was the case.

In providing electric service to urban areas, electric utilities took on a far greater risk than that posed by rural electrification. Before any revenue could be received, a distribution system had to be built with no assurances that its cost would be covered. Not only did customers have to acquire appliances, those appliances had to be invented and made available for sale. Buildings had to be wired. Encouraging greater use of electricity in urban areas was a constant concern, an area where Samuel Insull had been particularly successful. If providing rural service turned out badly, a utility's profits would have suffered. Providing urban service required accepting a risk that the firm would fail, figuratively "buying the farm." There were utility executives that realized this difference quite early, one of whom urged his colleagues to devote more effort to rural electrification:

> This business, however, is not going to come to us without our taking the initiative. The electric vehicle, the flatiron, the general power motor, and even the electric light did not come into their own without your having had the courage of your convictions and embarking a slight amount of capital before you had the revenue to be gained actually laid down across your counter. . . . your rural revenues are a great deal more certain to-day than were the revenues from flatirons, motors and lamps when you first took them up.[41]

The industry's willingness to accept risks was not confined to its early history. In the late 1920s, the privately owned utility industry was increasingly dominated by holding companies engaging in quite risky behavior. The negative consequences of those risks became manifest when the economy collapsed. What was there about rural electrification that prompted such timid behavior? Privately owned utilities maintained that economics could not justify rural electrification, only "sociology."[42] Farmers would receive electricity only if it were subsidized by the government or if regulatory agencies changed their policies. Yet there is no evidence that utilities were advocating such a regulatory change, and they opposed the uniform-rate proposal made in the Giant Power report. Nor was there indication that the regulatory commissions were interested in leading on this issue. As a result, the hostility of critics of the role of private ownership in the industry increased. The industry's position was destructive for the nation and ultimately contrary to its own self-interest. What can explain this situation? The most plausible explanation puts the blame on the structure of the industry that evolved from the system of state regulation. This explanation is supported by the impact that a new federal agency, the Rural Electrification Administration, had on both rural electrification in general and on the behavior of privately owned utilities specifically.

The Rural Electrification Administration

It is not surprising that the New Deal initiated a program to accelerate the pace of rural electrification. Although the proportion of farms with electric service had been steadily increasing, it remained low, and the industry offered no encouragement that its pace was going to increase. Farmers had long been a politically important group, and for a decade and a half their economic status had slipped compared to that of the rest of the country. Private ownership of electric utilities faced unprecedented attacks partially motivated by an ideological conviction that utility executives had sacrificed the public's interests in favor of their own. Those responsible for the New Deal's program to increase the availability of electricity to farms were not ideologues, and they did not treat the problem of rural electrification as another moral failure of privately owned utilities. Perhaps inevitably, rural electrification did create new friction with privately owned utilities. Their response shows that their self-interest had not been aligned with that of the public.

The industry's argument that its evidence and research had repeatedly shown that farm electrification could not pay its way was flawed, but the contrary position that it would pay its way took faith. If rural electrification were to become self-supporting, three things were required: long-term financing, low rates, and patience. A federal government agency brought all three. On May 11, 1935, President Roosevelt issued an executive order creating the Rural Electrification Administration (REA), and Congress gave the agency a statutory basis the following year.[43] Roosevelt appointed Morris L. Cooke as the first administrator of the REA. Cooke had been the director of the 1925 Giant Power Survey Board, and his appointment might have raised concern among privately owned utilities. However, Cooke was quite willing to work with privately owned utilities to achieve the REA's goals. Rural electrification had long been a concern of his, and after Roosevelt's first election, but before his inauguration, Cooke had urged him to initiate a program to bring electricity to between 50 and 75 percent of America's farmers.[44] Cooke recognized the difficulties faced by privately owned utilities, and he did not condemn them, but he thought the government might be able to help overcome those difficulties.[45]

The year before, the FTC had released its final report on the efforts of the industry to influence public opinion, and it was widely viewed as exposing malfeasance. The investigation's two other final reports were issued in 1935, and the Public Utility Holding Company Act was being considered. The TVA was being fought in the courts and in public opinion. The privately owned segment of the utility industry justifiably felt besieged. Two days before Roosevelt issued the order creating the REA, the president of the Edison Electric Institute told a conference of savings bankers that his industry had been "singled out for destruction" because "the President has an obsession on the subject."[46] Nevertheless, before the creation of the REA, Cooke held talks with industry executives and found them friendly and cooperative. He proposed that the administration create a joint program with the privately owned segment of the electric utility industry, but he encountered opposition from within the administration. When Cooke suggested this partnership to Harold Ickes, the secretary of the Interior, his response was, "I'll have nothing to do with the sons of bitches."[47] Nevertheless, Cooke continued to court the participation of privately owned utilities.

In 1935, the new REA was to be a funding agency, lending money for twenty years at 3 percent interest. Congress appropriated it $100 million ($1.7 billion in 2014 dollars) for this purpose. This addressed both financing the construction of a rural distribution infrastructure and giving farmers time to adjust to the use of electricity. The remaining requirement for rural electrification to succeed was low rates. Cooke wanted the REA's money to go to work immediately, and he realized that privately owned utilities already had the work force and expertise and would be the quickest way to create the new infrastructure. Two days after the creation of the REA, Cooke told the press that he had considered four different options for using the agency's money and that most feasible would be to lend the money to privately owned utilities. He also said that low rates were essential to the success of the plan and that if privately owned utilities were unwilling to provide them, he would be forced to use a different option.[48]

Cooke invited representatives of privately owned utilities to meet with him on May 20, nine days after the REA's creation. The meeting went well, and industry leaders promised to conduct a survey of the potential for rural electrification and appoint a committee of privately owned utilities to develop a plan for rural electrification. Voices within the industry predicted that a joint plan between the government and privately owned utilities would emerge and that rural electrification would occur under private ownership.[49]

At the same time, relations between the government and privately owned utilities continued to deteriorate, not because of the REA but because of the holding-company issue. On June 3, the annual three-day meeting of the Edison Electric Institute opened and provided a venue for privately owned utilities to vent their anger. The vice president and director of the EEI condemned the reports of the FTC as "fraud, deceit, misrepresentation, dishonesty, downright maliciousness, breach of trust," and asserted that the president and members of Congress had predetermined its findings.[50] In addition to repeating his characterization of Roosevelt as being obsessed, EEI's president attacked the federal government for "having for its object the end of private operation of the electrical industry and its nationalization under Federal direction and ultimate ownership." Other speakers said that the government's program was "heading down the same road as communism" and that privately owned utilities were "the victim of aggressive and well-paid propagandists for public ownership." The REA did not escape criticism at the convention. A member of the recently formed committee of privately owned utilities blamed political interference for having exaggerated the benefits of and minimized the cost of farm electrification. Farmers, he maintained, had greater use for things other than electricity.[51] Despite this rancor, on July 10, Cooke specifically asked for proposals on rural electrification from holding companies as well as operating companies. Stating that hundreds of applications for loans had already been received, Cook also said, "REA is dealing on equal terms with all groups willing to undertake to supply farmers with electricity."[52] A few days earlier, he had also invited cooperatives to apply.

It took almost two months for the chair of the committee of privately owned utilities to submit a letter to Cooke giving their plan.[53] The committee proposed that the REA lend the entire $100 million to the privately owned utilities in return for which the utilities would construct over seventy-eight thousand miles of new lines and connect 351,000 new customers, of which 247,000 would be farms. The cost to the utilities would exceed the amount loaned by just under $14 million, the source for which was not mentioned. Presumably, the utilities would cover that shortfall themselves. However, the committee warned that even with the government loans, it would be a long time before rural electrification could be self-supporting.

A number of points made in the letter must have disappointed Cooke. Since rural electrification would not be self-supporting, privately owned utilities continued to regard the program as a social, rather than economic, issue. Utilities would assume the sacrifice of carrying losses during a long development period, but the terms of the loans needed to be liberalized. Rather than twenty years, the term should be twenty-five years, or the repayment made a percentage of the revenue generated by the projects. The REA should realize that providing electricity to new farms would not result in new major use of electricity for farm operations because the farms that could make such use already were supplied with electricity. The most troubling assertion was probably that "the problem of the farmer was not one of rates." The farmer's problem, according to the letter, was covering the cost of residential wiring and purchasing enough household appliances to bring usage to the level that would justify being provided with electricity. The letter asserted that the average cost of wiring and appliances was $354, over seven times the average cost of a year's electricity. Therefore, in addition to the $100 million loaned to utilities, the government would also have to makes loans directly to the new customers for wiring and appliances amounting to over $124 million at low interest and under liberal terms. This was money that the new REA did not have and loans it did not have the authority to make. The letter stated that rate simplification would be a slow process because of regulatory constraints, and the companies were already using rates designed to encourage the greatest use of electricity. Nevertheless, despite the difficulties and financial risk, the committee said construction of new lines could begin immediately as soon as the new customers could be signed up "on the basis of the aid extended to them." The individual utility companies would have to make the final decisions, but the committee encouraged them to apply for loans.

In his reply, Cooke said he would refer the matter of appliance financing to the EHFA, but he disputed the committee's figure for the cost of wiring and appliances, maintaining that the average cost would be between $40 and $100, not $354. He strongly dissented from the stated industry position on rates: "On the contrary, we hold rate simplification and even rate reductions over large areas to be the heart of the problem of electrifying rural America." His letter was depicted as an acceptance, and he asked the committee to use its "good offices" to communicate with their utilities that the REA was prepared to accept applications and sought "maximum activity." Acceptance of the applications was not assured. "Naturally, in weighing the relative desirability of loans, it will be necessary for REA to consider carefully existing and proposed rate structures with reference to developing the large use essential to the success of our program."[54]

Within the privately owned utilities, suspiciousness and even hostility towards the REA grew. Caution in dealing with the REA was advised, and some held that the REA never intended to provide money to privately owned companies and was using those companies to gather information needed for providing funds to others.[55] The REA's initial proposals did include ones from privately owned utilities, and these were among the largest proposed projects.[56] The REA approved the proposed projects of seven private companies in 1935, but it notably rejected one from Wisconsin Power & and Light, whose president, Grover Neff, had long been seen as an advocate of rural electrification and was a member of the committee that had submitted the privately owned utilities' plan. Cooke rejected the proposal because he thought the company's proposed rates were too high. Neff charged Cooke with only being willing to extend loans to utilities that extended rates below the cost of service and urged a united front among the privately owned utilities in opposition to the REA. The effect of this was quickly felt. Although privately owned utilities in future years received a few loans, their share of the total plummeted (table 8.1). Cooke was also encountering opposition to cooperation with privately owned utilities from the other side. Senator Norris expressed to Cooke the concern that providing money to privately owned utilities would be tantamount to supporting the enemy.[57] Cooke's dream of cooperation with privately owned utilities was dead. Cooperatives became the primary recipients of REA loans. At the same time, privately owned utilities increased their own efforts at rural electrification by liberalizing their rules for rural extensions and building new lines. One factor that made it easier for privately owned utilities to reject the REA was that by that time those in good financial shape could obtain more attractive loans on the private market than those offered by the REA.[58]

TABLE 8.1 **REA loans by borrower type**

Year	Number of loans				Percentage		
	Total	Cooperative	Government owned (municipal, state, and PUD)	Privately owned	Cooperative	Government owned (municipal, state, and PUD)	Privately owned
1935	30	18	5	7	60.0%	16.7%	23.3%
1936	161	147	10	4	91.3%	6.2%	2.5%
1937	140	128	8	4	91.4%	5.7%	2.9%
1938	155	150	3	2	96.8%	1.9%	1.3%
1939	170	160	7	3	94.1%	4.1%	1.8%
1940	104	99	4	1	95.2%	3.8%	1.0%
Total	760	702	37	21	92.4%	4.9%	2.8%

Source: US Rural Electrification Administration 1966, 39–43

Congress gave the REA more permanent status in 1936 with passage of the Norris-Rayburn bill.[59] The new law confirmed and expanded the role of the REA and included some of the provisions of the privately owned industry's plan. Loans from the REA could only go to provide electric service to areas that did not already have service. Privately owned companies, state and territorial governments, municipalities, public utility districts, and cooperatives were all eligible to receive loans, but the law directed the REA to give preference to governments and cooperatives over private utilities. The loans were to be self-liquidating, could be made for a term of up to twenty-five years, and could be extended another five years. The interest rate charged for such loans was tied to the average rate paid by the federal government on bonds with maturities exceeding ten years. The REA was also authorized to extend a second class of loans to cover expenditures on wiring and on both electrical and plumbing appliances and equipment. These loans were not to go directly to individuals but to either the utilities that had borrowed funds from the REA or to those businesses selling or installing the wiring or equipment. The interest rate charged on these loans to the utilities was to be the same as that used to finance infrastructure. The term of the loan, however, was to be for only two-thirds of the life of the equipment financed or five years, whichever was shorter. The standard practice the REA adopted was to loan this money to cooperatives, which then relent it to individual members at 6 percent interest.[60]

The REA and Rural Cooperatives

Although the REA continued to make loans to government-owned and privately owned utilities, most of its work to increase rural electrification came through rural electric cooperatives, a policy continued by Cooke's successors. A cooperative is a nonprofit organization owned by those it serves. Final decisions are made democratically, with each member having one vote. Rural cooperatives had long been used in activities such as marketing, but, before 1930, few had been created to distribute electricity, and about half of those failed. A cooperative movement already existed in the United States, and both the proponents of cooperatives and farmers' organizations were advocating for cooperatives to play a major role in rural electrification. Rural electric cooperatives were widely, and successfully, used in Europe, and before the creation of the REA, the Federal Emergency Relief Agency had studied their use for rural electrification.[61]

The TVA was actively involved in a program of rural electrification mandated by the law that created it, and the TVA promoted cooperatives for rural distribution. Farmers in the TVA area were offered the same low basic rates as those in urban areas, although they also paid the amortization charge, making their monthly bills as much as $1 per month more than many urban areas. The TVA was working on reducing the cost of rural distribution lines, but, by 1935, farm electrification in its service area was still not widespread, and in its 1934 annual report, the TVA reported that low electricity use initially continued in some of the rural cooperatives it served.[62] Nevertheless, the rapid economic success of TVA's first rural cooperative in Alcorn County, Mississippi, was seen as establishing the viability of cooperatives as providers of rural electrification.[63]

Five years before the creation of the REA, there were only thirty-three electrical cooperatives in operation in the United States, and they were quite small. The number of customers served by these cooperatives ranged from four to 267 with a median of only thirty-five.[64] By the end of 1935, the TVA was providing power to five cooperatives, none of whom had (yet) received REA loans. They were much larger than earlier cooperatives; the number of customers ranged from 632 to 2,086 with a median of 868. There were other cooperatives formed between 1930 and 1935, but they were likely not many.[65] The policies of the REA greatly accelerated the formation of rural electric cooperatives, and by the end of 1940, the REA had made loans to over 700 cooperatives (table 8.1).

Setting up a rural electric cooperative involved technical, regulatory, and business issues that were likely beyond the capabilities of the farmers who wanted service. To aid the formation of cooperatives and to protect its investment, the REA became deeply involved in both the creation and the continued operation of the cooperatives that owed it money. Once it received a request from a group of farmers, the REA would send the farmers information, and, if their interest continued, a team of specialists would travel to the area to assist them and guide them through getting legal support, setting up an organization, obtaining any need regulatory approval, designing a system, and negotiating a wholesale contract. This included direct involvement when problems arose in dealing with a wholesale electricity supplier or a regulatory commission. If a cooperative had difficulty getting sufficient membership, the REA would become directly involved in persuading neighbors that they would benefit from the use of electricity, including using travelling demonstrations of household electrical appliances and farm equipment. By 1937, however, the demand for loans exceeded the available funds by $90 million, and there was no further need for development work.[66]

Before submitting a formal loan application, the REA required the cooperative to be legally incorporated and to have signed up a sufficiently large number of members, each of whom had paid a fee, usually $5 (approximately $84 in 2014 dollars). The REA's legal staff prepared most or all of the required legal documents. The cooperative's board chose a local attorney subject to rejection by the REA. A superintendent (or manager) and engineers for the project chosen by the board and approved by the REA had to be in place. Even after approval, the REA retained the power to dismiss both the project's superintendent and the cooperative's manager. The cooperative was required to have already prepared maps for the project and secured all needed rights-of-way previously checked and verified by REA staff. All required regulatory approval had to have been secured and a tentative wholesale contract negotiated.[67] Acceptance of a loan application by the REA depended on the REA's judgment as to whether the cooperative's project was on a sufficiently sound financial footing to expect it to be able to repay the loan. Many cooperatives were not successful in obtaining loans.[68]

Once the REA approved a loan and construction began, its involvement continued. REA approval was required before the cooperative could issue invitations to bid on the project, and an REA field engineer determined the winning bids. All contracts, including the wholesale power contract, required REA approval. The REA monitored the construction and could require that work on the project be redone if it judged the materials or the workmanship substandard. The construction techniques were developed or approved by the REA. The REA dictated the cooperative's accounting procedures, subjected it to periodic audits, and specified standards for the salaries of the cooperative's employees. Outside the TVA region, the REA effectively determined the retail rate schedules used by cooperatives. The support of these activities required a staff greater than that of just a lending agency, and between 1935 and 1939, that staff grew from 99 to 778.[69]

The REA was active in supporting electric cooperatives in other ways. Prior to 1935, only Iowa had a statute providing for the formation of cooperatives that distributed electricity. The legal staff of the REA developed expertise in the laws of all the states as they pertained to electric cooperatives. Many cooperatives were formed under general laws permitting nonprofit corporations that had primarily been used by charitable, religious, or educational organizations. Cooperatives in other states organized under general business incorporation statutes. There was considerable variation among the states, and the laws of some made the formation of an electric cooperative nearly impossible. Connecticut, for example, required an act of the legislature before the creation of any new electric concern. Electricity cooperatives faced legal problems unique to the distribution of electricity. These include the issues of access to the public rights-of-way, the relationship between electricity cooperatives and the states' utility regulatory commissions, and the extent to which an electricity cooperative had the same powers of eminent domain as a privately owned utility. In 1934, the Public Works Administration submitted a set of model state laws, including two concerning electric cooperatives, to all state governors. By 1936, six states had adopted such laws. In 1937, the REA drafted a model Rural Electrification Cooperative Act that exempted cooperatives from all state regulatory commission oversight and granted them full powers of eminent domain. The law encountered significant opposition in many states, but fifteen adopted it, although often in a modified form.[70]

The REA was also actively involved in finding ways to reduce the construction and operating costs of electric cooperatives. For example, the use of stronger lines permitted wider spacing of poles, and simpler poles and less expensive transformers and meters cut costs. Initially, these were not new techniques developed by the REA; private utilities and the TVA had already developed and used them:

Most of the actual construction techniques now being widely used were known two years ago and were being practiced here and there. In short, the role of the REA Engineering Division in the field of construction is not one of invention. It is one of gathering together the best practices known, coordinating them and stimulating their general adoption.[71]

The REA required all projects to have "area coverage," distribution lines built so that most farms within the project area would be able to connect to them even if they did not do so initially. Providing coverage to a large area enabled the REA to employ installation methods that took advantage of economies of scale by dividing the task among different specialized crews. Thus, one crew went where a line was to be constructed and marked the locations of the posts. Another crew followed the first, digging the postholes, another would place the poles, and still another would string the wires. This procedure contributed to substantial cost reductions. For 1935 and 1936, the total cost per mile of REA-financed rural lines was $1,085. In 1939, it was $700.[72] By contrast, the 1935 average cost for rural lines built by privately owned utilities was $1,252.[73] After their use by the REA, privately owned utilities adopted the same construction techniques. The REA worked with equipment manufacturers to reduce the costs of both the equipment used in the distribution system and that used in farmer's homes. This was aided by the development of uniform standards enabling multiple cooperatives to purchase jointly large amounts of the same equipment. The REA negotiated with insurance companies to reduce the premiums paid by cooperatives. The members of some cooperatives were given the opportunity to work on the projects themselves and were taught to do their own wiring to reduce costs further.[74]

Privately owned utilities were the major source of wholesale power for REA-financed cooperatives, providing almost 50 percent of all power used by the cooperatives during the twelve months ending June 1941. The purchase of wholesale power by a cooperative was its major single operating cost, on average constituting 39 percent of all operating costs and equaling 75 percent of the amortization and interest on its capital facilities.[75] The REA viewed securing a sufficiently low price for wholesale power as crucial to a cooperative's success; too high a price could prevent a cooperative from receiving REA financing.[76] One section of the REA, the Rate Section, had as its primary responsibility helping cooperatives negotiate wholesale rates.[77] When a successful wholesale rate could not be negotiated, the matter was taken to the courts, the legislature, or the state regulatory com-

mission, and these efforts were sometimes successful.[78] Congress had authorized the REA to loan money for the construction of power plants and
transmission lines as well as distribution systems. The REA's policy was
to do this only under one of three situations: (1) the cooperative's territory was too far from existing generating facilities, (2) existing generating
facilities were overloaded, or (3) no satisfactory wholesale rate could be
arranged. During the 1941 fiscal year, REA-financed generating plants accounted for only 4 percent of the total power supply used by all REA cooperatives.[79] Most plants were diesel powered, but some were gas powered
or hydroelectric. Usually these generating facilities were quite small, but
there were exceptions. Sometimes a group of cooperatives would join to
create a generating cooperative, a cooperative of cooperatives, that would
serve thousands of electricity users. The possibility of REA-financed generation could strengthen the negotiating power of a cooperative, and private utilities sometimes would reduce their wholesale price in the face of a
threat of an REA-financed generating plant.[80] The REA and the cooperatives argued that privately owned utilities should offer a special wholesale
rate available only to cooperatives that covered only the additional costs
to the utility of adding the cooperative as a wholesale customer to their
existing system, not the higher "full" costs generally used in rate making.
The privately owned utilities and state regulatory commissions sometimes
accepted this argument. When this resulted in a lower wholesale rate for
cooperatives than for other wholesale customers (including municipally
owned utilities), privately owned utilities sometimes faced discontent from
their other customers.[81]

The Reaction of Privately Owned Utilities

The law limited the REA to only loaning funds to provide electricity to
areas that were unserved, areas that the privately owned utilities regarded
as uneconomic. Furthermore, the REA was willing to make these loans to
utilities under private ownership. Unlike the TVA, the REA did not take
existing customers away from privately owned utilities. Instead, it offered
them prospects for new wholesale customers. Spokespersons for privately
owned utilities criticized the REA on the grounds that many cooperatives
would fail, creating a burden for taxpayers including privately owned utilities. Yet it is not clear that the failure of a cooperative would have been
harmful to a privately owned utility. If a cooperative defaulted on its REA

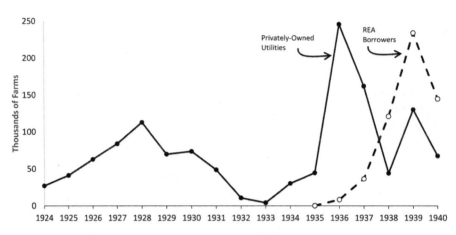

FIGURE 8.4. Annual number of farms first provided electricity by type of utility

Source: See note 82

loan, the REA would foreclose and take over the system. It might operate the system for a while, but it would eventually sell it. A likely buyer
would be the privately owned utility in whose territory the failed cooperative had operated. That privately owned utility would then have the opportunity to acquire a complete distribution system at a price below its
construction costs. Even if someone else, such as another cooperative, purchased the system, a privately owned utility would still have a wholesale
customer. It is very unlikely that the financial failure of a cooperative would
result in its electricity system being dismantled or becoming unused. In
addition, widespread failures of REA-financed cooperatives would be
evidence that the position of privately owned utilities about the financial
feasibility of rural electrification had been correct all along, perhaps providing schadenfreude at a time when they felt persecuted by the federal
government.

Cooke initially expected privately owned utilities to be the REA's primary borrowers, and they always remained eligible for loans. With few
exceptions, however, privately owned utilities had decided not to apply
for REA loans. Nevertheless, once the REA was established, privately
owned utilities began their own rural electrification on an unprecedented
scale (figure 8.4).[82] In 1936, privately owned utilities connected to their
systems five and half times as many farms as they had in 1935, a number
of farms equal to 45.6 percent of all of the farms they had connected the

previous decade. The rate dropped somewhat after that, but the annual average number of farms connected each year between 1936 and 1940 by privately owned utilities was about 130,000, 15 percent higher than had been connected in 1928, the previous peak, and 2.4 times the average of the previous decade. Privately owned utilities served about 45 percent of all farms first receiving electricity in 1940.[83]

With the sudden increased activity of privately owned companies, it is not surprising that there were territorial clashes. Many privately owned utilities significantly lowered the costs a farmer had to pay to get service, including eliminating the up-front connection charge, and the timing of these changes sometimes coincided with the formation of a cooperative in the area.[84] Privately owned utilities had an important advantage over cooperatives: they could move quickly. In some states, cooperatives did not have the power of eminent domain. For a power line to cross private property, the cooperative had to obtain an easement cleared by the REA from every property owner. By June 30, 1936, cooperatives had submitted more than two hundred thousand easements to the REA for clearance. By contrast, a privately owned power company did have the power of eminent domain and could use it to begin immediately installing lines.[85] It could take a year or more between when a cooperative began soliciting customers and when those customers were actually receiving service. Under these circumstances, it was not surprising that some farmers might have preferred certain and quick service from a private utility over future, and perhaps uncertain, service from a cooperative. There was justification for farmers' initial concerns. The pre-REA cooperatives did not have a good record, and the privately owned segment of the utility industry emphasized the risk of depending on a cooperative for electricity. In 1936, an executive of a major holding company described the US experience with electric cooperatives in these terms:

> . . . there are many advocates of rural cooperative lines, owned and operated by farm customers. This movement is not a new one, having been carried on for many years, in many States, and under varying conditions. We have had sufficient experience by this time to know the many weaknesses of such enterprises. Most attempts to furnish electric service to farms at lower rates than those of the private utilities have ended in failure and in the instances where they have been lower it has only been through the failure to take into consideration all of the costs of rendering service. Usually maintenance is so neglected that most of these lines are in a deplorable condition. . . . They are frequently turned over outright to the private utilities for maintenance and operation, or are sold, at

a fraction of their original cost. Hundreds of typical cases could be cited to demonstrate such results over the entire country. . . . The mere fact that the Government is financially and otherwise encouraging such enterprises is no indication that they will be more successful than in the past.[86]

Privately owned utilities sometimes discouraged farmers from joining cooperatives by using dishonest assertions. Farmers were told they would have to mortgage their farms to receive service from a cooperative, that they would be personally responsible for the cooperative's debt to the government, that they would be personally responsible for maintaining the cooperative's lines in front of their property, or that they would have to replace transformers that burned out. Some farmers heard that they would be liable for any accidents to third parties caused by the cooperative's equipment. Farmers were also assured that their bills would be higher than if they took service from the privately owned utilities.[87]

Solicitors from both a cooperative and a privately owned utility sometimes poured into the same area. Some farmers signed contracts with both. The financial success of a cooperative could be threatened by the loss to a privately owned utility of those customers in its project area who would use the most electricity. A privately owned utility had a legal obligation to serve anyone in its territory willing to meet its conditions, and it likely considerably liberalized those conditions. In some states, these territorial disputes ended up as headaches for regulatory commissions. The Wisconsin commission once stated that it needed "virtually the powers of a mind reader to ascertain which service is actually desired."[88] Privately owned utilities were not above taking customers from a cooperative by making deceptive promises. In one case, the REA accused Duke Power of having promised immediate service to residents of one rural area if they would cancel their agreement with a cooperative. Duke Power claimed that all of the county's residents wanted out of the cooperative, and the cooperative's entire project was dropped. The company then failed to provide the promised service for months, and eventually the cooperative submitted a new loan proposal to the REA.[89]

The REA asserted that often what was at stake in the territorial disputes was area coverage. The projects of cooperatives were required to provide access to most farmers in the area. The past practice of many of the privately owned utilities was to offer service only to those for whom it would be most profitable. On many occasions, the REA asserted that by serving only the best customers in an area, a privately owned utility had doomed a cooperative's project, preventing most of those in its project area from getting service from anyone. Sometimes, but not always, regulatory commissions and courts agreed.[90] The mechanism privately owned companies were accused of using to cripple or destroy a cooperative included "cream skimming," "blocking," and "spite lines." "Cream skimming" referred to the practice of taking the cooperative's best potential customers. "Blocking" referred to a privately owned utility preempting a strategic right-of-way, preventing a cooperative from extending its lines. A "spite line," whose name reveals the emotions involved, was a line constructed by a privately owned utility into the area of a cooperative for the express purpose of cream skimming or blocking. Privately owned utilities that had previously refused to construct power lines without receiving payment in advance were accused, once an REA loan was approved, of hurriedly constructing miles of spite lines in the middle of the night, under floodlights, without having signed a single customer.[91] The director of the REA asserted that spite lines had affected 200 cooperatives in thirty-eight states, that eight cooperatives had been forced to close, and that over one hundred thousand consumers consequently were denied service. Among the many individual cases he cited was one where a privately owned utility used a regulatory commission hearing to delay for months a cooperative from beginning operation, during which time the utility built 150 miles of spite lines in the center of the cooperative's best territory.[92]

The REA forbade the cooperatives it financed from providing service in areas where it was already available but decided this did not prevent a cooperative from building lines parallel to those of a privately owned utility if the latter's construction had started after the REA had approved the cooperative's project.[93] In a number of states, regulatory commissions restricted privately owned utilities from extending service into territories claimed by cooperatives by requiring commission approval for rural extensions, or by insisting the utility negotiate with the cooperatives, requirements that likely reduced the rate at which privately owned utilities connected farms.[94] The REA claimed it had no problem with a privately owned utility that promised to provide area coverage to a rural area, even if a cooperative project had begun, if that was the wish of the people in the area. Loans were rescinded in these cases, and, in at least one case, a portion of the area to be served by a cooperative was removed from a project after a privately owned utility agreed to provide that portion with service.

Privately owned utilities sometimes used their role of supplier of wholesale power to hamper the development or operation of a cooperative. Some privately owned utilities simply refused to provide wholesale electricity to cooperatives at any price. Some contracts they offered had onerous clauses. For example, before providing power, some privately owned companies required cooperatives to agree that none of the electricity they distributed would be sold at a lower rate than that of the privately owned utility.[95] The REA frequently maintained that the wholesale prices private companies quoted cooperatives were excessive. The rates of any electric utility are difficult to evaluate, but when the Kentucky Public Service Commission investigated the question of wholesale schedules used by privately owned utilities for cooperatives, six of the seven companies declined to provide any factual cost basis, and the commission regarded the submission of the one other company as inadequate.[96]

It is impossible to verify the extent to which the friction between privately owned utilities and cooperatives were innocent consequences of both trying to serve the same rural customers or the product of deliberate efforts to sabotage a cooperative. This was a time when ideological polarization and the repressed desire of many farmers for electric service created an emotional atmosphere in which the activities of privately owned utilities could have been misinterpreted. However, James Bonbright, perhaps the era's preeminent scholar on the economics of electric utilities, had this to say on the issue in 1940:

> Recently the Rural Electrification Administrator has complained that several companies, anticipating plans to electrify a given area by means of a farm cooperative, have tried to defeat these plans by a hasty construction of "spite lines"—lines that invade the richer portions of the area without serving more than a small fraction of the territory. The companies accused of this practice have denied the charge vigorously. But the complaint comes from public officials of such high integrity that it cannot fairly be dismissed without better evidence than the corporate executives have yet brought forward.[97]

In each of its earlier annual reports, the REA indicated that opposition from privately owned electric utilities had diminished. Nevertheless, this mention shows that it had not disappeared, and the 1940 report indicated that in the late spring of that year there had been a marked increase in spite-line activity. Opposition to REA-financed cooperatives seemed to occur only at the individual-company level. There was not the same type of organized industry-wide opposition against cooperatives or the REA as had been seen with the TVA. There were a number of court cases brought by privately owned utilities against cooperatives, but these primarily involved local issues. There were no basic constitutional challenges to the REA program. Alabama Power did seek an injunction in a federal district court that would have brought a halt to the REA lending program, but the injunction was denied, and Alabama Power did not seek an appeal.[98]

There is no doubt that electrification of farms had a major impact on the quality of life of a significant proportion of the population, many of whom were very poor, but there also is evidence that electrification very quickly had a positive economic impact on farm operations as well. A recent analysis of the impact of the REA found evidence that by 1940 electrification increased the value of farm output, retail sales, and land and even reduced infant mortality.[99] The REA was only five years old in 1940, and most of the cooperatives that depended on it were younger. The work of electrifying farms had not finished by 1940. It was not until the mid-1950s that the level of electrification of farms equaled that of the suburbs, and it was 1975 before 99 percent of US farms had electricity.[100]

Did Farm Electrification Require the REA?

The REA gave rural electric cooperatives a new major role in the industry, but their inclusion was not, by itself, a major change in the industry's structure. Distribution was then still primarily done by integrated utilities also involved with generation and transmission. By 1935, however, there were also many utilities, both privately owned and government owned, that only distributed electricity to a local area. These distribution-only utilities were integral to the TVA and the later Bonneville Power Authority systems. The REA, however, did bring a short-term disruption to the structure of the industry by creating a competitor to privately owned utilities for distributing electricity in rural areas. The establishment of the REA coincided with a dramatic change in the rate of farm electrification (figure 8.1). In 1935, only 12.6 percent of the nation's farms were provided electric service. In 1940, the percentage had more than doubled to 32.6 percent, much of which was the result of increased activity by privately owned utilities. Should the REA be given credit for the changed behavior of privately owned utilities?

If the privately owned utilities were correct that rural electrification was not viable, how was the REA able to stimulate the creation of large numbers of financially viable cooperatives? The REA's low-cost construction

techniques certainly helped, but this explains neither the behavior of privately owned utilities before the REA nor the dramatic change in that behavior after. The largest privately owned holding company systems each
dwarfed the entire set of projects funded by the REA prior to 1940. Many
of the construction techniques used by the REA had been developed by,
and used by, privately owned utilities. Any of those holding company systems, or the entire industry through the NELA, could have mounted a
program to lower the costs of rural distribution on a far larger scale than
that of the REA. Privately owned systems could have been at least as
effective as the REA in encouraging manufacturers to produce new and
less expensive farm equipment and household appliances. The REA combined cooperative purchasing to negotiate lower prices, but this had long
been a standard practice of holding companies. The REA's success cannot
be ascribed to an inherent technical advantage it possessed over privately
owned utilities. Before the REA, the interest of privately owned utilities
in rural electrification was low, but the creation of the REA made it much
more attractive.

Might the REA's program have provided a hidden subsidy to rural electrification? Such a subsidy could have enabled a rural cooperative to serve
an area where an unsubsidized private utility would have lost money. There
was no overt subsidy; the REA made loans to cooperatives that were required to repay those loans with interest. Nevertheless, the REA's program
provided an implicit subsidy in two ways. The loans made by the REA
to cooperatives were at an interest rate generally below that paid by privately owned utilities that borrowed money through the sale of securities.
In addition, the federal government paid the entire administrative costs
of the REA from general revenue; they were not repaid by the loan recipients. This contrasts with the situation of privately owned utilities that
paid the administrative costs, such as the fees of investment bankers, of
the loans they received. Furthermore, the administrative costs of the REA
went beyond those of a mere lender and included a high level of technical
and administrative support privately owned utilities paid for themselves.

Before the REA, privately owned utilities usually required farmers to
pay in advance for the cost of the service extension they required. Although
there was a wide variation in these charges, a common figure for the cost
of a line was $1,200 with a typical density of three farms per mile. This im-
plies an average cost of $400 per farm. By contrast, a reasonable estimate
for the value of the REA subsidy per farm in 1940, using the interest rate
privately owned utilities were paying, was only $65 (table 8.2 and this chap-

ter's appendix). Had the value of the subsidy been closer to $400, it might have explained how the REA was successful when privately owned utilities were not. Even then, it could not explain why, even without the subsidy, privately owned utilities suddenly became so willing to provide rural electrification.

Another possible reason cooperatives may have been able to succeed where privately owned utilities could not was that cooperatives enjoyed preferential tax treatment. As with government owned utilities, this is difficult to quantify. It requires deciding what aspects of the tax treatments of cooperatives constituted an implicit subsidy. There was great variation among states in the ways they taxed rural electric cooperatives. Most states did not exempt cooperatives from the payment of taxes, although at least one state, Mississippi, provided such an exemption in 1940. The REA did not advocate tax exemptions for cooperatives, except, perhaps, during a cooperative's initial years, but it did advocate a ceiling on state taxes based on a cooperative's gross revenues. The "area-wide" requirement that cooperatives provide access even to farms that did not take service resulted in low revenues compared to the value of the cooperative's property. The REA reported that state taxes sometimes amounted to 25 percent of gross revenue. Some states modified their taxes for cooperatives, reducing their tax liabilities. Minnesota, for example, taxed cooperatives based on their number of members, initially 10¢ per member. The REA argued that all taxes should be limited by a taxpayer's ability to pay, which they noted would be very different for a developing cooperative than for an established privately owned utility serving a more profitable area.[101] Despite the REA's advocacy, it had no power over state governments and state taxes. The motivation for state governments to extend tax breaks to cooperatives may have been to promote rural electrification. Had privately owned utilities actively lobbied at the state level for more favorable tax treatment for such projects, they might have been successful, but there is no evidence they tried. Rural cooperatives were nonprofit and thus paid neither state nor federal income taxes. The value of other special tax treatments from the federal government was slight. Cooperatives were exempt from a 3 percent federal energy excise tax and a few other minor federal taxes.

The experience of both rural cooperatives and rural electrification projects by privately owned utilities confirmed that, contrary to earlier assertions, rural electrification would pay for itself. Prior to the REA, aversion to the risk of extending service to rural areas was deeply imbedded in privately owned utilities. The slow start of the REA in 1935 could not have changed expectations about the likelihood that rural electrification would prove uneconomic. Nevertheless, 1936 saw an explosion in rural electrification by privately owned utilities. Why?

Industry Structure, Regulation, the REA, and Incentives

State regulation protected electric utilities from competition in their service areas, and most of the nation's farms were in the service area of a regulated utility. That utility had a legal obligation to provide service to anyone, including farmers, who were willing to meet the conditions established by the company and approved by the regulatory commission. Those conditions were set so high that few in rural areas could meet them. This protection from competition meant that for a utility, the decision was whether to provide service to a rural area now or later. This choice discouraged taking risks. When utilities were first providing service in urban areas, they did not yet have guaranteed territorial monopolies. A utility that hesitated to provide service to an area risked losing that area to another utility, and utilities were willing to take considerable risks to provide service. At the same time that they were hesitant about rural electrification, holding companies were taking risks to acquire control over operating companies in competition with other holding companies.

The REA stripped away much of the protection state regulation provided to privately owned utilities in rural areas. Instead of having the choice to postpone providing service, a utility that failed to serve a rural area might forever lose that area to a cooperative. If a privately owned utility believed that providing service to a particular area would never pay, the threat from a cooperative would have had no effect. If the utility saw new rural service as providing a long run positive benefit, it had an incentive immediately to make the needed investment. Before the REA, a long-run positive benefit was not sufficient. The benefit had to be realized in the short run. If the choice were now or later, the time to make the investment was when its payoff would have the largest present value. Rural electrification was widely expected to come eventually. Over time, the use of electricity in farms would increase, and farms that first received service then would begin using larger amounts of electricity than farms first receiving services now. As urban areas expanded, the distance to some farms would decrease, making it less expensive to extend service to them.

In this situation, alternative investments might seem more important. This was exactly the claim made in the 1922 report by the NELA's Rural Line Committee.[102] Cooperatives brought other factors to consider. If a utility allowed a cooperative to operate, there was a good chance it would get a new wholesale customer. It would lose, however, the ability to price discriminate among users in the area. The electricity sold at wholesale would not have as high a potential profitability as electricity sold at retail. If there were a nearby federal hydroelectric facility, the cooperative would get preferential access over the privately owned utility to the cheap power it could provide. The REA and the threat it brought from a cooperative completely changed the way a privately owned utility evaluated the economics of rural electrification. That change resulted in evaluations of rural electrification and alternative investments more in line with the public interest. The REA corrected a flaw in the industry's incentive structure that had been created by state regulation. That flaw only involved unserved areas, and the corrective function of cooperatives stopped once electricity became available in all areas.

The view that the nation's experience with rural electrification was bound to the structure created by state regulation is a logical hypothesis. Perhaps future research will be able to test it with empirical evidence. The protection state regulation provided to privately owned utilities covering unserved areas varied at least somewhat among the states. In some states, public utility districts might have weakened that protection, and states varied in the ability of municipally owned utilities to extend their service beyond the municipality's boundaries. These variations might offer an opportunity to examine the extent to which differences in rural electrification were due to differences in state regulation's territorial protections.

The creation of the REA was a notable example of a government policy having a positive effect on the electric utility industry. The REA's program did include a modest subsidy. From the perspective of federal taxpayers, the value of the subsidy incurred to provide electricity to a farm, rural residence, or business in 1940 for twenty-five years was $36, or $601 in 2014 dollars. The benefit of that subsidy extended beyond the individuals who received service to include those whose service was provided by the privately owned utilities whose policies were changed by the REA. Although there was initial speculation that with REA support, privately owned utilities would be the engines of rural electrification, it is not surprising that did not happen. The poisoned political climate was certainly a factor inhibiting privately owned utilities from taking greater advantage of REA support, but supporting privately owned utilities was not a solution to the problem of rural electrification. The solution was to bring competition to the provision of rural electric service. The competitive threat rural cooperatives brought hastened the availability of electricity to all Americans. They no longer have that role to play, but the prominent part rural cooperatives continue to play in the US electric utility industry is a legacy of the benefit they brought the industry and the nation.

Appendix to Chapter 8: Calculation of the Implicit Subsidy of the REA's Program

The value of the REA's implicit subsidy depends on the value of an interest rate different from that used by the REA. The choice of rate depends on the perspective from which the subsidy is to be evaluated. Its value to a cooperative's membership, the privately owned electric utilities, and federal taxpayers differed. To value the subsidy from the perspective of taxpayers, the interest rate paid on federal debt should be used. To value it from the perspective of privately owned utilities, the rates they paid to obtain funds should be used. Finally, to value the subsidy from the perspective of the farmers and others who received REA-sponsored service, the correct interest rate is the one they would have paid to obtain the financing needed to construct a distribution system without the REA. Table 8.2 shows the calculations using all three of these rates. These subsidies are per REA customer (meter), such as a household or business. They are the lump sum present value in 1940 of the REA's subsidies for twenty-five years, the term of loans issued in that year.

TABLE 8.2 **Implicit one-time subsidy per 1940 customer of REA cooperative**

Comparison group	Average interest rate, 1936–1940	Implicit subsidy of REA loan	Implicit subsidy of REA administration	Total subsidy, 1940 dollars	Total subsidy, 2014 dollars
Farmers	6.00%	$88.70	$33.53	$122.24	$2,050.17
Private utilities	3.53%	$23.81	$40.84	$64.65	$1,084.27
Federal government	2.58%	-$8.61	$44.45	$35.83	$600.98

Source: See chapter appendix

By 1940, the REA's program was sufficiently underway that reasonable data are available for most of the required measures. That date is early enough that an estimate of the implied subsidy would be relevant for evaluating the positions and actions of the privately owned utilities because the ultimate success of the program was not yet known. The calculations assume that the information about interest rates and administrative costs known in 1940 would persist for the next twenty-five years. Both changed, but to evaluate the implications of the subsidy in 1940 (and earlier), only information available then was used.

There are two parts to the implicit subsidy of the REA's program. The subsidy of the loan is the value in 1940 of a cooperative borrowing money at REA interest instead of an alternative rate. The higher the alternative interest rate, the greater that portion of the subsidy. The other implicit subsidy comes from the REA's administrative costs, which were not charged to those served by the REA. Those administrative costs were annual; a higher alternative interest rate makes their present value in 1940 lower.

The average interest rate charged by the REA between 1936 and 1940 was set by the average rate paid by long-term federal government bonds, 2.81 percent, and this was used in the calculations. The REA's rate changed considerably after 1940. In 1945, Congress set the rate at a flat 2 percent. Although reasonable at first, alternative rates rose significantly, increasing the total implicit subsidy of the REA's program. The gap did not become apparent until after 1950, but it continued to grow for decades.[103] By this time, the work of the REA was nearly complete; 78 percent of farms were receiving electrical service.

The rate for farmers is based on a statement made by Harry Slattery, the REA administrator, that farmers could earn 6 percent on their money.[104] The relevant rate, however, is not the rate farmers could receive but the rate they would have had to pay a willing lender for a twenty-five-year loan on the terms provided by the REA. Undoubtedly, that would have been higher than 6 percent, increasing the value of the implicit subsidy. Even likelier, such a loan would not have been available at any interest rate. Generally, a private lender will require collateral whose value exceeds the amount of the loan. The REA accepted as collateral the value of a system built with its loan and the $5 fee paid by a cooperative's members. If a cooperative were unsuccessful, that collateral would have been worth less than the loan. Perhaps rural residents could have mortgaged other assets to satisfy a private lender, but that would not have been easily arranged. If no alternative financing could have been obtained, the value of the REA's implicit subsidy to a customer without service from an electric utility is both greater and more difficult to calculate. It might be approximated by the additional cost required to provide equivalent service from an isolated plants or an estimate of the total consumer surplus that utility service would have provided. $88.70 is a lower bound, perhaps far below the true value of the REA's program to farmers.

The alternative interest rate used for private utilities is based on the rates paid on long-term utility bonds and preferred stock. Between 1936 and 1940, the average interest rate paid by privately owned utilities on long-term bonds was 3.44 percent.[105] Data on the preferred stock dividend rates paid by privately owned utilities were published by the *Journal of Land and Public Utility Economics* only for 1939 and 1940. For those years, the yield on preferred stock was 1.22 percentage points above that for long-term bonds, and that difference was assumed for the other years. The weighted average rate from both financial instruments is 3.53 percent. These rates exclude the fees and other administrative costs privately owned utilities incurred in issuing these securities. The equivalent costs for the REA's loans were part of the agency's administrative costs and are handled in that portion of the implicit subsidy. No adjustment was made for common stock or short-term debt. Data on the proportion of money raised from common stock are not available, determining an appropriate value to use for a return is problematic, and the great majority of operating company common stock was held by holding companies. The proportion of total funds that came from holding company common stock sold to outside investors was very small.

The federal government rate is the average rate for all federal debt at the time. Since that rate includes short-term debt, it is lower than the REA's rate, which was based only on long-term bond rates. Ideally, a risk premium should be added to this rate. Had an REA borrower defaulted, the REA would eventually have taken over the system and sold it. Defaults would have occurred when a cooperative's revenue was insufficient to cover loan payments. The eventual sale of the system by the REA would probably not have covered the full amount of the loan, and taxpayers would have made up the loss. To the REA's detractors, this was a likely outcome.[106] Some of the first REA cooperatives initially encountered financial difficulties. The REA placed a portion of the blame on the activities of privately owned utilities and liberalized loan terms.[107] In 1940, no loan had been outstanding for more than five years, most would have another two decades or more to run, and many borrowers had yet to make a first payment. There was no objective basis for assigning a risk premium in 1940. In June of 1941, only 1.2 percent of all payments were in arrears. Payments received by the REA in advance of their due dates were over twenty-five times that amount. In 1964, all of the loans made prior to 1940 would have been either due or almost due, and more loans would have been made. Of all of those loans, only two ended in foreclosure, with a total loss to the federal government of $44,478.[108] The risk of REA loans proved negligible and no risk premium was attached to the federal government interest rate.

The average value of an REA loan per customer was $315 by 1940 (about $5,300 in 2014 dollars).[109] The interest and principal was assumed amortized by twenty-five annual payments of $17.33. The implicit subsidy of the loan is the difference between the $315 from the REA and the amount that could have been borrowed at an alternative rate for twenty-five annual payments of $17.33. Since the federal government rate was less than the REA's, that implicit subsidy is negative.

The REA's administrative expenses covered the many services provided rural cooperatives and the costs of administering the loan program. Congressional appropriations for these in 1940 were $2,700,000.[110] The average for all customers for that year was $3.19 ($53.50 in 2014 dollars). It would have been reasonable to expect these per-customer costs to decrease over time. Initially, as the number of projects increased, scale economies would reduce the cost per project, as had already happened. Between 1937 and 1940, the REA reported that the costs per project fell from $9,000 to $1,714.[111] The REA's involvement in a project would decrease once the construction of the project was completed. In 1967, the REA reported its administrative costs per customer as $2.39 in 1967 dollars, or 92¢ in 1940 dollars. Beginning in 1952, the REA had the added responsibility for telephone service in rural areas, and it reported that 46.5 percent of all person-hours in 1967 was devoted to telephones rather than electricity; 92¢ overstates the administrative costs in that year for just the electricity program.[112] The REA's administrative costs were assumed to have declined linearly between 1940 and 1967 from $3.19 to 92¢, and the present value of those costs in 1940 is the estimate of that portion of the implicit subsidy.

Conclusion and a Look Forward From 1940

Before the nineteenth century, electricity was a scientific, and some-times popular, curiosity. The nineteenth century brought numerous technical advances, and during the first half of the century, these included commercial uses in chemical processes and in the transmission of messages through electric telegraphy. In the last quarter of the century, the development of arc lighting brought a practical and improved form of artificial lighting, albeit with limited application. Thomas Edison revolutionized the commercialization of electricity with a new system. He invented neither the electric light nor the incandescent bulb, but he developed the system that brought electricity into homes and offices and laid the basis for the modern electric utility industry. The unique characteristics of his incandescent bulb lay at the center of his new system, but the radical new system required solutions to problems never previously appreciated. His genius was in anticipating those problems and inventing the solutions that made the system workable. Since then, the impact of electricity on modern life has been profound. Its effect on manufacturing productivity improved the nation's standard of living. The huge infrastructure created by the utility industry supported (and continues to support) the development of additional innovations, including those responsible for the current information technology revolution. Once available, electricity changed from being a luxury to almost a necessity of life, one that could not be interrupted even briefly.

The value of electricity for some purposes is so large that even extremely high prices would not end its use. At the prices actually charged for electricity, its net benefit is enormous. Before 1940, increases in the price of electricity were very unusual and decreases were common. One might suppose that an industry with this record would have brought to its customers satisfaction not controversy. Instead, the electric utility industry was chronically the subject of political controversy to an extent seldom experienced by other industries.

The structure of an industry determines the incentives facing the industry's decision makers and the extent to which their interests align with those of the public. For most industries, public policy has sought to encourage a structure that comes close to one where many firms compete with one another and new firms can freely enter the industry. The economics of electric utilities, however, makes this ideal inappropriate and does not provide a clear alternative. Disagreement over what would be the best alternative structure has been the basis of all the industry's controversies. Until World War II, or shortly after, the disagreement centered on whether the nation was better served by utilities owned and operated by government or by profit-seeking privately owned firms. The electric utility industry is unusual in that it has always included both.

Electric utilities have a unique set of economic and technical characteristics that have limited the set of workable industry structures. With minor exception, electricity must be produced in the instant it is used. Unlike other networks, the path taken by the flow of electricity from a particular producer to user cannot be controlled or identified. Decentralized control of a network is impossible. At the same time, a single fully integrated network can serve a geographic area more economically than multiple independent networks. These characteristics make competition effectively impossible. A very large proportion of the costs of an electric utility, particularly the early utilities, was the cost of equipment, and a large proportion of that equipment was transaction-specific. Once it was acquired, it had little value if it could not be used for its original purpose. The combination of the need for a utility to be the sole provider in a particular area and the importance of transaction-specific capital equipment resulted in widespread municipal corruption when individual firms had to depend on franchises to operate.

A recognition that monopoly supply of electricity was needed joined the realization that monopoly power could be used to exploit customers and discourage the efficient use of electricity. These led to the development of state regulatory commissions that had authority over the prices charged by privately owned utilities. This type of regulatory commission had been tried with railroads but was unsuccessful until it was used with utilities. By then, the Supreme Court had laid down conditions that regulated prices had to meet. Those conditions were ambiguous and confusing and hampered the implementation of regulation. Those who were suspicious of profit-seeking privately owned utilities saw regulation as ineffective and continued to advocate for government ownership. They were joined by conservationists whose concerns centered on the exploitation of natural resources. That, and the growing importance of the interstate sale of electricity, drew the federal government into the industry both as regulator and as participant.

Regulation provided a durable industry structure within which the industry grew and made enormous contributions to the nation's economy and well-being. That structure, however, was deeply flawed, and those flaws resulted in many new policy proposals. Regulation created disincentives for utilities to operate as efficiently as possible. Barriers created by regulations discouraged the development of large integrated networks, particularly interstate networks. World War I brought awareness of this deficiency and led to two proposals that would have addressed this problem and restructured the industry. Neither, however, was enacted.

The importance of capital equipment made electric utilities heavily dependent on outside investors. Holding companies facilitated access to those investors and provided the early utilities with technical and administrative support. Holding companies were also able to exploit weaknesses in the methodology of state regulation, enabling utilities to regain much of the monopoly power regulatory commissions were supposed to deny them. However, holding companies were also able to overcome the barriers state regulation posed to the development of large integrated networks, and they built a number of such networks that continue in operation. An intention to create larger networks cannot explain the scramble by holding companies to acquire far-flung operating companies, particularly during the last half of the 1920s. During the 1920s, holding companies enjoyed spectacular financial success and received considerable interest from Wall Street. The growing financial and political power of holding companies led to multiple federal investigations. The most exhaustive federal investigation of any industry concluded that holding companies had used tactics to deceive investors and had exploited electricity users by taking advantages of weaknesses in state regulation. The investigation was also very critical of state policies, including those used by many state regulatory commissions.

The stock market crash and Great Depression dramatically changed the nation's political climate and the public's view of the privately owned segment of the electric utility industry, especially holding companies. A retrograde policy, the Public Utility Holding Company Act of 1935, sought to return the industry to a structure that had existed before holding companies became dominant. The law forced the breakup of nearly all holding companies and made the formation of new holding companies unattractive. Many states reformed their regulatory commissions. The barriers to new large integrated networks that holding companies had been able to overcome were re-erected.

New arguments in the New Deal for government ownership and operation of electric utilities held that private ownership should continue to dominate the industry but that a limited role for government would be a more effective check on the undesirable behavior of privately owned utilities than state regulation. The federal government assumed a significant role as owner and operator of electric utilities. The most dramatic instance of this role was the Tennessee Valley Authority, which took on all of the responsibilities of an electric utility as part of a much larger mandate. It was a government agency with a unique organization. The advocates of government ownership hoped the TVA would become a template for similar authorities across the country, but privately owned utilities saw it as an existential threat. The TVA was able to create a large fully integrated network, but its organization was not replicated, and its continuing influence on the industry eventually ended. The federal government's role as a producer of hydroelectricity, however, continued to expand.

Perhaps the most successful federal program that identified and corrected a flaw with the structure of state regulation was the Rural Electrification Administration. State regulation had reduced the incentives for privately owned utilities to extend service to rural areas. The REA enabled the creation of a large number of cooperative distributors, but an even more notable effect was the change it caused in the behavior of privately owned electric utilities, which began providing service to farms at an unprecedented rate. The protected monopoly status of electric utilities had discouraged their providing that service earlier. The REA changed that behavior by bringing competition to the rural distribution of electricity.

Electric Utilities after 1940[1]

By 1940, the changes in the industry structure brought by New Deal programs were not complete. The breakup of holding companies continued into the next decade, new hydroelectric sites were developed by the federal government, and electricity had not yet been made available in all rural areas. During World War II, the production of electricity by all sectors of the industry grew rapidly to meet war needs. Soon after the end of the war, the impacts of New Deal programs had largely been completed. The boundaries of the TVA were set, eliminating its continuing threat to privately owned utilities. The federal government's role as a provider of hydroelectricity ceased growing once all feasible hydroelectric sites had been developed. The breakup of holding companies ended, and virtually all rural areas were provided access to electricity. Advocacy for government ownership of utilities remained a local issue in some areas but disappeared as a national issue. A 1944 Supreme Court decision on the power of regulatory commissions eliminated much of the confusion caused by the court's earlier decisions, and reformed state regulation provided the dominant structure for the industry. Its flaws largely remained, and the industry structure it created continued to inhibit the development of new large integrated networks, but the demand for electricity grew rapidly, and the industry entered a three-decade-long period of relative quiet, during which the use of electricity grew at an annualized rate of over 8 percent.[2]

A confluence of factors brought to an end the industry's period of political peace in the 1970s. The 1973 Arab oil embargo made "energy crisis" prominent in the nation's political lexicon, and oil use became seen as a vulnerability that threatened national security. This occurred after a period in which environmental concerns had resulted in federal policies encouraging utilities to shift from coal- to oil-fired generation.

Optimistic projections of future growth in electricity usage led utilities to plan major construction projects. After a long period in which the federal government had promoted the use of nuclear energy without much success, the electric utility industry enthusiastically embraced the new technology and began undertaking planning and construction of power plants whose scale far exceeded those of prior plants. Nuclear power offered low operating costs but high capital costs that became much greater than the industry's initial expectations. A period of rapid inflation brought soaring nominal interest rates at a time when the industry was trying to raise the substantial funds required to build the new plants. Electricity prices rose faster than the rate of inflation but not fast enough to keep up with costs, and utility profits fell. Growth in the demand for electricity fell short of the industry's forecasts, and utilities were forced to cancel projects underway on which they had already spent enormous sums.[3] State regulatory commissions faced the problem of finding a way to pay these costs. Utilities argued that their customers should pay the costs based on a controversial argument for the existence of a "regulatory compact." They maintained that regulators had approved in advance the utility's failed construction plans, and regulation restrained the rate of return a utility could earn on successful investments. Since regulation prevented utilities from earning high returns when investments performed better than expected, it also limited the losses to a utility when investments turned out badly. For decades, the battle over exactly who would pay these sunk costs would consume enormous effort and resources.[4]

This experience with failed utility investment decisions revealed a new flaw with the regulatory system. If an investment decision by an unregulated firm in a competitive environment turns out badly, it is the firm, not its customers, that bears the consequences. A risky investment, however, that turns out well may temporarily provide such a firm a large return. Faced with these possible outcomes, a firm considering undertaking a risky investment has an incentive to devote great care to investment planning and to forecasting future demand. One aspect of utility investment planning was the choice of technology to use in future plants. Of all that were available, nuclear power posed special financial risks. Even if the total cost of generation from that technology was expected to be cheaper than alternatives, careful planning required those risks to be taken into account. For example, an alternative with very different financial and economic characteristics was gas-fired turbines. The cost of constructing a nuclear plant, its capital costs, far exceeded those of a gas-fired plant, but its operating costs were then lower. The total cost of electricity from a nuclear plant might have been expected to be lower, but this depended on some critical assumptions about the reliability of a technology with which the industry had little experience. Because of the high fixed capital and low operating costs, the total cost of electricity from a nuclear plant is little different whether or not it is actually generating electricity. To be a cost-effective way of producing electricity, a nuclear plant must run as close to continuously as possible. At least partly because of the industry's lack of experience, nuclear plants experienced many unforeseen events that forced them to be shut down, and the cost of the electricity they generated was consequently high.

Other factors made nuclear technology financially risky. Forecasts of future demand are uncertain. If the actual demand for electricity falls short of those forecasts (as happened), losses will be lower if the plan was to meet that unrealized demand with a lower capital-cost technology. Forecasted demand should not be the sole basis for investment planning. Confidence in those forecasts, and the consequences of their being inaccurate, must also be considered. The longer expected construction times of a nuclear plant contributed to its financial risk because forecasting errors tend to increase over time.

The issues associated with investment risk do not mean that risky investments should never be undertaken, but their consideration mandates devoting more care and more resources to both investment planning and demand forecasts. Luck remains a major factor in determining the outcome, but careful planning can reduce the consequences of bad luck. A criticism of regulation is that it reduced the incentive for electric utilities to devote the care and resources that were needed. Regulatory commissions do not experience the full consequences of investment losses, and they do not have the resources needed to plan effectively for risky investments. Requiring a utility to obtain the permission of a regulatory commission prior to undertaking investments was not a solution to the problem of investment risk. Shifting the consequences of poor investment decisions to customers reduces the incentive for those making the decisions to devote the care and resources investment planning requires. Yet this was explicitly done by many state commissions, who adopted "construction work in progress" policies, which immediately added construction costs to the rate base instead of waiting until the plant was operating. This enabled a utility to receive a portion of the costs (interest costs) of a cancelled plant even before its cancellation. Firms operating in a competitive market cannot similarly charge current customers for investments that will benefit only future customers.

The clarity of hindsight made it apparent that utilities had not done a good job of investment planning, and the incentives created by the system of state regulation were at least partly to blame. One way of eliminating this problem was to restructure the industry so that firms planning for investments in generation operated in a competitive market rather than as regulated monopolists. Other developments further encouraged the shift of generation from regulated monopolists to competing firms. In 1977, numerous provisions in a series of laws that came from President Carter's national energy plan affected the electric utility industry. Utilities had complained that delays in licensing nuclear plants contributed to their long completion times, and a new law streamlined that process. Other laws addressed problems not directly related to the nuclear imbroglio. Environmental concerns had caused utilities to shift generation from coal to oil and natural gas. The new law required them to bear the costs to shift generation back to coal. Recognition that existing rate structures discouraged the most efficient use of electricity resulted in a requirement that state regulatory commissions consider better rate structures, including rates based on the time of day of electricity use. The laws also created a new cabinet-level department, the Department of Energy, and the federal regulation of utilities that had been the responsibility of the FPC was transferred to the new Federal Energy Regulatory Commission (FERC).

The law that had the largest impact on policies affecting the industry's structure was the Public Utility Regulatory Policies Act. This law required utilities to purchase electricity from independent firms (termed "qualifying facilities," or QFs) that produced electricity either from renewable resources or as a by-product of other industrial processes (cogeneration). The state regulatory commissions determined the prices utilities were required to pay based not on what it cost to produce the electricity but on the utility's "avoided cost," what it would have cost the utility to produce the electricity it was required to purchase. Regulatory commissions might have interpreted avoided cost as just the additional cost a utility would have incurred had it produced the electricity instead of purchasing it from a QF, its marginal cost. That interpretation would not have included in the price paid a QF any of the costs of underutilized power plants, since those costs were the same whether or not the plants were actually used for generation. Such prices would have encouraged generation from the most efficient source, but those prices would have been low. A number of commissions saw benefit in encouraging the development of the seemingly environmentally friendly new generation sources and set prices much higher. The high prices proved very profitable to the independent QFs and they became a significant source of all new generation. Many utilities entered into long-term contracts to purchase this expensive electricity.[5]

Generation from outside the regulated utilities was further encouraged by the Energy Policy Act of 1992, which created a new class of independent generators, "exempt wholesale generators" (EWGs). These were not restricted in the method used to produce electricity, but negotiation, not regulation, determined the prices they received. Regulated utilities were compelled to use their systems to transmit the electricity produced by the EWGs to their wholesale customers. Falling natural gas prices made gas-fired generation attractive, and EWGs were able to rapidly construct the inexpensive plants that used this technology. EWGs had a major advantage over the regulated utilities. Major components of the high prices charged by regulated utilities were the costs of unneeded and unreliable nuclear plants and the high prices they were forced to pay QFs, often under long-term contract. The bargain prices charged by EWGs created a rush by wholesale customers (including municipally owned distribution systems and rural cooperatives) to shift their source of electricity from the regulated utilities to these new generators, and their numbers grew. The savings enjoyed by the customers of EWGs made them vocal advocates for more independent generation and access to utility-owned transmission lines. Industrial users that were not wholesale customers clamored to get access to the electricity generated by independents and advocated "retail wheeling" so they too could get access to the cheaper electricity. Some firms, including regulated utilities, hesitated to create independent generating subsidiaries for fear they would become subject to the regulations the SEC imposed on public utility holding companies. Congress responded to their concerns in 2005 by repealing the 1935 Public Utility Holding Company Act. Since the customers of the independent generators were not contributing to the costs of nuclear plants and high-priced power from QFs, their growing popularity contributed to the problem of recovering those costs, which became known as "stranded costs." The political problem of paying these stranded costs increased.

At the time all of this was happening, price regulation, similar to that used with electric utilities, was being eliminated from other industries, including natural gas production, airlines, and trucking. If there ever was a rationale for regulating these industries in ways that set prices and limited competition, it was no longer apparent. Deregulating those industries caused considerable turmoil among the affected firms and did result in changes to their structure, but public policy did not attempt to dictate or guide those changes. Technological changes in telecommunications made restructuring of the telephone industry inevitable, although the existing telephone industry resisted such change. Restructuring of that industry, including separation of the sale of telephone equipment from the provision of telephone service and separation of long-distance from local service, occurred under the direction of federal courts and proved more complicated than the deregulation of other industries. Nevertheless, the experience with all of these industries showed that the elimination of price regulation was feasible.

The British experience with their electric utility industry was an important precursor to developments in the United States. In 1926, Britain's development of the national grid created a structure in which generation and transmission were under separate ownership. Generation was not deregulated, but some competitive behavior did emerge.[6] This model had been considered before in the United States—with the TVA power pool proposal and the Magnusson proposal at the creation of the BPS, and it was inherent in the Giant Power proposal. Britain nationalized its electric utility industry after World War II but privatized it in 1989 under a new structure. That new structure heavily influenced the development of the new electric utility industry structure that emerged in parts of the United States.[7]

Different areas of the United States implemented versions of the new structure. Although they differ in details, they are very similar in important ways. A regional transmission organization (RTO) has the responsibility for operating the transmission system, dispatching generating plants, balancing generation and use, and maintaining the overall stability and reliability of the entire electricity distribution system.[8] RTOs do not own the transmission network or significant amounts of any of the assets that make up the electric supply infrastructure. They neither generate electricity nor distribute it to final users. RTOs are nonprofit organizations subject to FERC regulation. They sell wholesale electricity to separate organizations (privately or municipally owned distributors or cooperatives) that distribute and sell electricity to retail users. An RTO organizes and operates markets where it purchases electricity from unregulated competing generating firms at prices determined by a market.

Each day the generating firms supply the RTO with bids for each hour of the following day, specifying the generating capacity they are willing to make available and the minimum price they would accept for the electricity they generate. The next day, the RTO dispatches generators in order of their offer prices, lowest to highest. The highest offer price of any generating firm actually selling electricity to the RTO is the price paid all firms during that hour.[9] The prices generating firms receive thus change each hour and those prices determine the prices the RTO charges distributors. In addition, the RTO runs a real-time market to purchase electricity required to maintain the system's balance. There are other specialized markets, the need for some of which was realized only with experience. For example, in order to ensure that generation and use are always perfectly matched, standby generators capable of beginning generation instantly are required. To eliminate the delay required to bring turbines and generators up to speed, these generators are kept running even when they are not producing electricity. Their owners incur costs and require compensation even if the RTO does not purchase electricity they produce. RTOs operate separate markets where firms compete to provide this service.

A number of states have adopted a version of the new structure, including Texas, California, New York, and several states in New England, the Midwest, and the mid-Atlantic. Other states, including most in the Southeast, have retained the traditional system of state regulation of integrated utilities. In the 1990s, it appeared that the entire nation was in transition to the new system, but a crisis in California in 2000 and 2001 halted that process. A large share of the effort that went into the development of California's structure was devoted to the issue of recovering stranded costs, and this likely contributed to flaws in the structure the state implemented.[10] Users and suppliers of electricity were prohibited from entering into long-term contracts, and retail prices were capped in many areas. A number of factors, including reduced rainfall and transmission bottlenecks, lowered the available supply of electricity in several areas. The unregulated generating firms were able to exercise market power by withholding their electricity even when the prices they received would have covered their costs. In these circumstances, the decision of a single firm could drive the price everyone received above that firm's marginal cost. Withholding some electricity reduced the amount it sold, but the higher price more than made up for that loss. Since many areas capped retail prices, the higher wholesale prices did not lead to reduced use. Wholesale prices soared and the RTO was unable to acquire sufficient electricity to maintain system balance. Rotating blackouts in large areas were necessary, and one old utility, Pacific Gas & Electric, became bankrupt. Ultimately, the state entered into long-term contracts at high prices to ease the shortages.[11]

After the California experience, states that had not yet adopted the new system abandoned plans to do so. Some of the states retaining the system of regulation of integrated utilities, however, allowed existing RTOs to take control of their transmission systems. As a result, the multistate transmission systems under unified control became larger. There is evidence that these RTOs have been able to capture some of the benefits of large fully integrated networks.[12] Since they do not own the transmission system, the RTOs do not have the same control over transmission investments, as would a fully integrated utility that both owned and operated the transmission system.

The separation of transmission from generation adversely affects decisions to make new investments in transmission. A fully integrated utility has an incentive to bring electricity to its customers at the lowest possible cost and to choose the best mix of generation and transmission to achieve that goal. No similar incentive generally exists in a system where generation and transmission investment decision are made separately. In addition, as explained in the introduction, it is difficult to compensate an owner of a transmission line for the contribution that line provides the network because the flow of electricity between a single generator and a single user cannot be traced. The addition of a new transmission lines affects the entire network in ways that could be positive or even negative. Investment planning for the transmission system must be done centrally. Such planning requires forecasts of the level and pattern of electricity usage. Once such forecasts are made, the design of the network is a tractable engineering problem, but the economic incentive to construct a network that will supply future electricity needs in the most efficient way is muted. Ironically, the effort to correct a flaw in state regulation and improve decisions about investment in generation may have exacerbated a similar problem with investment decisions in transmission. Such a system can work, but neither those planning for new transmission investments nor those actually making the investments fully bear the risk of those investments. The system lacks the direct economic incentives present when an integrated utility owns and controls both transmission and investment.[13]

A further irony has arisen from a shift in the favored technology for new generating plants. The problems state regulation created in the incentives for utilities to engage in efficient investment planning became manifest because they were planning for huge power plants that a used new technology, that took a long time to complete, and whose costs were dominated by construction costs. By contrast, the technology that now dominates investment planning is natural-gas-fired plants. These tend to be smaller, they use a technology similar to that long used by the industry, and their costs are not as heavily weighted by construction costs. Had that technology been dominant in the 1970s, there might have been no stranded costs and no public policy initiative to restructure the industry.

Public policy initiatives that seek to increase the proportion of generation from wind and solar power increase the importance of a robust transmission system. Generators powered by wind or directly by the sun have a characteristic similar to that of hydropowered generators; the amount of electricity that can be generated at a particular time depends on the whims of nature rather than the need for the electricity. Compared to hydroelectricity, the availability of electricity from these two sources is even less predictable and more variable. The importance of robust networks in areas that depended primarily on hydroelectricity historically resulted in those areas having the best-developed transmission networks. If wind and solar are to become more important sources of electricity generation, increased attention to transmission networks will be necessary.

Although it now seems impossible to use market competition to develop the nation's transmission network, some states, notably Texas, have brought such competition to some aspects of electricity retailing. In these states, individual users of electricity can choose among different companies to supply them with electricity. These companies generally neither own nor control any of the physical assets involved with electricity generation, transmission, or distribution. They enter into contracts with generating companies, and they collect payment from their retail customers. Since all of these retail companies use the same distribution system, customers of the different companies must pay the same distribution costs. Moreover, since they all use the same distribution system, none has any control over the reliability of their customer's service. There is no difference in the electricity available from any of the suppliers, and they are sharply limited in the ways they can differentiate their services. By contracting with different generators, they can vary in the extent to which the electricity they sell comes from "green" sources. They can also vary in the type of financial arrangements they offer customers, with some specializing in customers with poor credit records. A new opportunity for differentiating will occur when the owners of distribution systems switch from the old kilowatt and demand-charge meters to newer "smart" meters. With these meters, competing retailers will be able to use different rate structures, including rate structures that better reflect the cost of producing electricity and better promote its efficient use. Competition in this area may encourage the adoption of better rate structures.[14]

Is the new industry structure that eliminates integrated utilities and brings competition to generation better than the older system of state regulation?[15] The answer is unclear. The incentives to efficiently plan for and operate generating facilities are better under the new system, and there is evidence that it has resulted in slight improvements in the operation of generating plants. There is stronger evidence that the use of RTOs to control large multistate transmission networks has brought positive benefits. States that have chosen to retain the system of regulating integrated utilities, however, can enjoy that benefit. The new system may adversely affect transmission planning and investment, but no study has analyzed current data to support or refute that concern. Fully integrated utilities may be better able to capture the full benefits of large fully integrated networks, and recent mergers following the repeal of the 1935 Holding Company Act increased the size of the networks under the control of such utilities.[16] Although those networks unify operation and investment planning, they remain smaller than the ones under the control of RTOs.

Electricity prices under the new structure are more volatile than those under traditional regulation. Futures markets have arisen in response to this volatility to allow electricity users to protect themselves against that volatility. If prices in futures markets influence investment and electricity use, they can contribute to greater efficiency. However, these markets and the specialized firms that operate within them use resources not used where traditional regulation is used, and this is a cost of the new system. There is evidence that during times when supplies are tight, individual firms will possess market power, enabling them to withhold generation resulting in a price above marginal cost and exploiting customers much as a monopolist could.[17] This is surprising for a market containing many sellers each providing an identical product, but market power explains many of the price increases in California in the summers of 2000 and 2001, and the special characteristics of generation markets may make this problem endemic.[18] There is also a plausible case that prices under the new structure could fall so low that generating firms would be unable to stay in business, but there is no evidence this has happened or of the likelihood of it happening.[19]

The comparative prices charged users under the two systems will drive the popular and political evaluation of the two systems. Those prices have been higher in states that adopted the new structure than in those that retained the old system. The two groups of states, however, differ in the many important factors that affect the price of electricity. The choice between the two industry structures likely has a much smaller effect on retail prices than do these other factors. Nevertheless, the lower prices in those states with the traditional system of regulation make it unlikely they will opt to change.

Across its history, there have been a number of major changes in the structure of the electric utility industry. The adoption of state regulation replaced the system of municipal franchises. Holding companies arose, aided the financing of the industry, built multistate integrated networks, and at least partially nullified many aspects of state regulation. The dismantling of the holding company system empowered a regulatory system that added significant federal regulation to that exercised by the states. Most recently, the development of RTOs with control over larger networks seems to have had a positive effect on network operations, although its effect on network investment and network development is uncertain. The new system has both advantages and disadvantages over the old, and the lack of clarity over which dominates has left the nation with two very different industry structures operating in different regions. Throughout the industry's history, flaws in the industry's structure led to policy-induced changes. The new structures that resulted brought new flaws that were usually unanticipated. Those new flaws have always created uncertainty as to whether the new structure was an improvement over the old, but those flaws encouraged new policy proposals to again alter the industry's structure.

The basic economic characteristics of the generation and distribution of electricity have remained unchanged, they make efforts to reform the industry's structure difficult, and they have made the design of a best industry structure elusive. There were numerous unadopted proposals for altering the industry's structure. These included the Superpower proposal, the Giant Power proposal, several power pool proposals, proposals for a vastly greater role by government-owned firms, and a proposal for electricity networks to be designed and implemented by a government regulatory agency. The authors of the Superpower proposal solved the problem of a new technical design for a greatly improved electric utility infrastructure in an important region of the country, and they even developed a workable technical transition plan. The new infrastructure, however, required a new industry structure without which the technical plan was impossible. Despite the huge unquestioned benefits the new technical structure would have brought, designing a new structure of ownership and control and a transition to that new structure proved intractable. Technical problems have always been easier to overcome than economic ones.

The ideological divisions that characterized so many of the battles over the industry's structure during its first half century have abated, and that has enabled the industry to move out of the center of public concern for long periods. A crisis that suddenly increased the price of electricity, as occurred in the 1970s, or reduced its reliability, as occurred in California in 2000, would again make the industry and its institutional structure a matter of prominent political controversy. Electricity is too important, and the problem of getting right the structure of the industry that produces it is too difficult, to imagine that its days of political controversy are over. We live in a time when many look to technological breakthroughs for solutions to problems. Is there a technological innovation that might eliminate the industry's cycle of political crises and restructuring by public policy? Such a change would have to alter the industry's basic economic characteristics. Perhaps the development of a device that was inexpensive, that could store large amounts of electricity, and that was small enough to be used by a household could do this. Such a device would profoundly alter the need for generating capacity and the economics of the transmission and distribution of electricity. If this were combined with a new technology that economically enabled dispersed generation, the impact on the electric utility industry would be disruptive. Pending those developments, the electric utility industry will remain an essential part of the nation's economy and infrastructure supplying, without interruption, a necessity of life.

Notes

Unless otherwise noted, all price conversions were made using the Bureau of Labor Statistics' Consumer Price Index, and calculated annual growth rates for multiple years are annual rates with continuous compounding (average differences in annual logs).

Introduction

1. "Nation to Be Dark One Minute Tonight after Edison Burial" (1931).

2. At the end of the twentieth century, some parts of the country restructured their electric utility industry in a way that enabled competition in generation but not transmission. Local distribution systems retain monopoly characteristics, although in some areas customers can choose among competing retailers for billing. This is discussed more in the conclusion.

3. For alternating current, inductive loads can result in a current's voltage and amperage being out of phase. Equipment designed to handle this type of current is often sized by the maximum volt-amps (VA), a measure of apparent, as opposed to real, power. Volt-amps and watts are always the same in direct current.

4. Oliver Williamson has applied the concept in this way in several works, including Williamson 1979; Williamson 1985; and Williamson 1988. It has also been applied in the context of regulated utilities; see Goldberg 1976.

5. Consider this simplified explanation. Suppose use of electricity on a network is constant at all times. The probability of a single generator failing is 10 percent; network reliability is 90 percent. Suppose a second generator is put into reserve. It will be operated the 10 percent of the time the first one fails. Since it will fail 10 percent of the time it is operating, there is a 1 percent probability both generators will fail at the same time; the network is 99 percent reliable, and its reserve capacity is 100 percent of its need. If a network needs ten generators, 99 percent reliability will require four reserve generators, only 40 percent of need. If the network needs 100 generators, 99 percent reliability will require 20 reserve generators,

20 percent of need. Larger networks will need fewer generators and a lower proportion of reserve generators than would be required by multiple smaller networks serving the same area.

Chapter 1

1. There is some debate over whether von Guericke's device was an electrostatic generator or the inspiration for an electrostatic generator developed by others. Schiffer 2003, 18–20.

2. US Bureau of the Census 1905, 86.

3. King 1962, 385.

4. Bowers 1998, 1–3.

5. Ibid., 20–40.

6. In 1885, after commercialization of both arc and incandescent lights was underway, Carl Auer von Welsbach invented the gas mantle, consisting of a cotton mesh impregnated with metal salts. When the mantle was first heated, the cotton would burn away, leaving the metal oxides. When heated, the metal oxides would incandesce, producing a light that, though harsh, was comparable in brightness and quality to an electric light, and the amount of light produced was over six times that of a gas flame. Its commercialization in the United States in 1890 retarded the adoption of electric lights. There were districts in New York in the 1890s where more Welsbach mantles were installed than Edison incandescent lights. "The Welsbach Light" (1900); Neil 1942; Passer 1953, 196–97.

7. The light of an open flame comes from the incandescence of incompletely burned material in the flame. Virtually all artificial light prior to electricity was produced by incandescence. Some methods of producing light from electricity, including gas discharge, do not use incandescence. These other methods are usually more efficient; more of the energy is converted to light rather than heat.

8. King 1962, 392.

9. Ibid., 396, 404.

10. Passer 1953, 13.

11. Among the others were Hiram Maxim, St. George Lane-Fox, Moses Farmer, William Sawyer, and Albion Man. Bowers 1998, 99–100.

12. Bowers 1982, 120.

13. Friedel, Israel, and Finn 1986, 14, 64, 66, 120, 177. While developing his electrical system Edison made well-researched and detailed cost comparisons with gas lighting systems to be as certain as possible that not only would the electric system be competitive but also that he would make a profit. Friedel, Israel, and Finn. 1986, 66–67, 120–22.

14. Schroeder 1986, 526.

15. Passer 1953, 99–102.

Chapter 2

1. The modern private electric utility industry prefers the term "investor-owned" to "private." Government-owned utilities often prefer the term "public" ownership to "government" ownership. "Public" when referring to ownership can be ambiguous since, as a rule, modern utilities are public corporations, that is, corporations whose stock is available for purchase by the public. The early census reports used the terms "commercial" and "municipal," but electric utilities came also to be owned and operated by states and by the federal government. A third class of electric utility that became important in the 1930s is nonprofit cooperatives, where the customers directly own the utility without an intervening government. Legally, cooperatives are privately owned, but when the industry was divided into two political camps, cooperatives were grouped with government-owned utilities rather than privately owned for-profit utilities. In many countries, hybrid ownership has also been common— corporations whose stock is held both by governments and by private parties. This has been very rare in the United States. The two forms of ownership will generally be designated here as "privately owned" or "government owned." Cooperative ownership will be discussed in chapter 8 with rural electrification.

2. Carlson 1991, 300–301, 360–61.

3. "J. Pierpont Morgan" 2014.

4. Passer 1953, 12–19.

5. Carlson 1991, 255–57.

6. Passer 1953, 118.

7. Ibid., 108.

8. Ibid., 140.

9. Ibid., 34–40.

10. Ibid., 37, 99.

11. McCormick 1981, 256, 258–59.

12. These deficiencies were acknowledged in the report. US Bureau of the Census 1905, 7.

13. Ibid., 7-8.

14. Based on the number of utilities under each ownership type in 1902, about 4 percent of each group switched ownership. The growth in the total number of private utilities during this period was slower than that of municipal utilities, not because private utilities had become less important but because they had experienced a degree of consolidation that had not happened with municipal utilities. US Bureau of the Census 1910, 20, 28.

15. Dorau 1929, 2–3.

16. Hausman and Neufeld 1991.

17. This point was made by Sam Peltzman (1971). He developed a theoretical model comparing the pricing behavior of a municipally owned utility with a profit constraint with that of a profit-maximizing privately owned utility. The municipal

utility could accept a lower profit that enabled it to offer lower prices to all users, but residential users would get the largest reduction. He then compared the prices of government-owned and privately owned utilities in 1966 but failed to find evidence that residential users benefitted under government ownership. One reason he offers for this is that, in 1966, privately owned utility rates were subject to state regulation. If these political considerations played a role in the decision to have municipal versus private ownership, it seems less likely that they were important in the establishment of the first municipal utilities when private ownership was not an option. The earliest available data permitting a broad comparison of prices charged by utilities under the two forms of ownership is for 1935. At that time, residential users served by municipal firms paid higher prices than those served by private firms, a result that may have reflected the scale economies the much larger private firms then enjoyed. Earlier studies, including those for prices in the late nineteenth century, tended to compare the costs to a city of street lighting from a city-owned plant versus a privately owned plant. The results were mixed. Hausman and Neufeld 1994.

18. Anyone who has used jumper cables to start a dead car battery for a gasoline engine knows that those cables are much thicker than the ones required for a home appliance, even an electric heater. The current from a car battery has only twelve volts; high amperage (and thick wires) is a requirement for high power. The power required by many home appliances is greater than that used by a car, but the 110 (or 220) volts in household current means that lower amperage (and smaller wires) can carry the higher-power current.

19. Today, the long-distance transmission of current commonly uses current at hundreds of thousands of volts. It took some time before technology made these large voltages available to electric utilities.

20. Denayrouze and Jablochkoff 1877.

21. Edison 1883. Ironically, modern technology has enabled the transformation of direct current, and high voltage direct current has advantages over alternating current for transmission. Alternating current still dominates transmission, in part because of the costs of conversion between alternating and direct current required by the switch made by the early industry.

22. Hughes 1983, 86–91.

23. Apparently, Gaulard and Gibbs did not fully appreciate the versatility that came from varying the number of turns in the coils. Their system, at least in part, lowered voltage by connecting multiple primaries in series and having multiple secondaries in parallel. Passer 1953, 132–35.

24. Parallel work was also done in Europe. See Hughes 1983, 95–98.

25. Garcke 1911.

26. Hughes 1983, 108; Neil 1942, 332.

27. Hughes 1958.

28. Most historians who have examined this period have been quite critical of

Edison's apparent efforts to slow progress (e.g., Hughes 1983, 107). For a contrary position and overview of the event see David 1991.

29. One obvious use for motor-generators or rotary converters was as direct-current transformers where a direct-current motor operating at one voltage drove a direct-current generator producing another. This approach was used in England initially in 1892 and was known as the "Oxford system." Direct-current generators produced 1,000-volt current that was reduced to 110 volts using rotary converters. Bowers 1982, 143–44. This approach would, presumably, have been limited by the difficulty of designing motors or generators that could operate at high voltage. By 1911, however, motor-generators were used in which the motor side ran directly off (alternating current) voltages as high as 15,000 volts. Gear and Williams 1911, 49.

30. The rotary converter was between four and eight percentage points more efficient at converting twenty-five cycle AC to DC than was a motor-converter using a synchronous motor, and its initial cost was also somewhat less. A table comparing the efficiency and initial cost of motor generators and rotary converters under a variety of conditions is reproduced in Gear and Williams 1911, 45. It is credited to a paper presented by Allen before the Association of Edison Illuminating Companies in 1908.

31. Gear and Williams 1911, 26–27.

32. Paul David and Thomas Hughes asserted that rotary converters quickly allowed an interface between the "old" DC technology and the "new" universal AC system, but this was not the case until at least the second decade of the twentieth century, and complete integration came much more slowly, as shown by the persistence of 25 Hz AC (see below). David and Bunn 1988, 181–82; Hughes 1983, 121, 208. Rotary converters that converted 60 Hz AC to DC did exist, but motor-generators were more commonly used for this conversion. Gear and Williams 1911, 44–45. However, a case of rotary converters used to convert 60 Hz AC to DC in Cleveland is given by the 1912 Census. US Bureau of the Census 1915, 113.

33. Gear and Williams 1911, 38–43. Frequency conversion by motor-generator is also described in US Bureau of the Census 1910, 101.

34. Murray et al. 1921, 152. This report, discussed in more detail in chapters 4 and 8, advocated a shift to 60 Hz for the region between Washington and Boston. For that entire region in 1919, 45.5 percent of generation was 25 Hz, and 46.8 percent was 60 Hz.

35. Hausman and Neufeld 1992, 23. This is measured by the ratio of total generating capacity to the total capacity of a utility's connected load. Surprisingly, most of the utilities (all but those that were 100 percent AC) had more total generating capacity than required by all the connected lights and motors. Perhaps the excess capacity was used for backup.

36. US Bureau of the Census 1910, 16. Capacity of DC constant voltage generators increased 23 percent, but this was offset by a 44 percent decrease in the capacity of DC constant amperage generators.

37. US Bureau of the Census 1910, 101.

38. At the 1899 meeting of the Association of Edison Illuminating Companies a paper and discussion concerned the question of the relative advantages of distributing current to consumers as alternating or direct current, although there was little question that the new generating plants would produce alternating current. It is clear that many of the new customers of these companies serving large cities would receive electricity that had been converted to direct current. There was considerable discussion of the problems posed by the need to use 25 Hz AC generators to supply those ultimately using direct current and 60 Hz generators for those ultimately using alternating current. The discussion of Commonwealth Edison's system in Chicago makes clear that a complete separation existed then between the system that ultimately distributed direct current and that distributing alternating current. Wagner 1900.

39. Consolidated Edison 2007.

40. Insull and Chicago feature prominently in Thomas Hughes's work (1983, 201–26). A more recent work on Chicago's electrification is Platt 1991. Insull's reputation suffered enormously during the Depression but has been considerably rehabilitated by Forrest McDonald's biography (1962). Insull is discussed in later chapters, particularly 5 and 6.

41. Junkersfeld 1903.

42. The introduction has a fuller discussion of this aspect of electric utility economics.

43. Neil 1942, 330.

44. Sweall 1913, 621, 645.

45. US Commissioner of Labor 1900, 551.

46. Ibid., 548.

47. Analysis of the profitability of electric utilities in the US Commissioner of Labor's report indicated that electric utilities just before the turn of the century were providing their investors returns no better than could have been received with far safer railroad bonds. Hausman and Neufeld 1990.

48. Layson 1994. This requires that average costs decline as output increases (increasing returns to scale), which was very likely the case for most electric utilities. If price discrimination resulted in higher sales, savings from reduced costs could have benefited everyone. Internal discussions among utility employees showed an awareness of this possibility. Sweall 1913, 656.

49. For example, the first block might have been 100 kWh at a price of 6¢ per kilowatt-hour per month. The second block might have been an additional 100 kWh, but at a price of 5¢ per kilowatt-hour per month. A third block might also have been for an additional 100 kWh at a price of 3¢ per kWh. A customer who used 100 kWh or less in a month would have paid 6¢ per kWh for all electricity used that month. A customer who used 150 kWh would have paid $8.50—$6.00 for the first 100 kWh at 6¢ per kWh and 5¢ for the next 50 kWh at 5¢ per kWh. The average

kWh price to that customer would have been 5.7¢, but an additional kWh would have cost only an additional 5¢. A customer who used 250 kWh in a month would have paid $12.50—$6.00 for the first 100 kWh at 6¢ per kWh, $5.00 for the next 100 kWh at 5¢, and $1.50 for the last 50 kWh at 3¢. The average kWh price for this customer would have been 5¢, but any kWh increase in use would have cost only 3¢.

50. The proposition that price discrimination can yield higher profits when lower prices are charged those with higher price elasticity of demand can be found in most, if not all, microeconomics texts and was certainly known early in the twentieth century, if not earlier. See, for example, Taussig 1926, 205. It is similarly well known that if prices on average must exceed marginal costs to cover total costs, greater economic efficiency occurs when different prices are similarly based on differences in demand elasticity. The first to derive this finding applied it to taxes. Ramsey 1927.

51. This slightly simplifies the factors determining the production cost to the utility of the electricity served a particular customer. Large utilities often used older, less efficient plants for use during times when usage was high but less than the peak. The additional cost of providing electricity from these "shoulder" plants was higher than that from the more efficient "base load" plants. The important point is that the production cost of providing electricity to a customer at a particular point in time was determined by the pattern of usage for the entire system, not the pattern of that customer.

52. The use of isolated plants was mentioned in the previous chapter. More discussion on their importance is in the next chapter.

53. This discussion of the demand charge rate structure as an instrument of price discrimination comes from Neufeld 1987. The discussion of the historical debate within the industry on rate structures comes from Hausman and Neufeld 1984.

54. Kapp 1892; Oxley 1897.

55. "Comments on Barstow" (1898, 133); Wright 1896; Wright 1897, 159–89.

56. "Report of the Rate Research Committee" (1914). The report includes a general discussion of price discrimination. Such rates are termed as being based on "value of service" as opposed to "cost of service," which must also figure into the design of rates.

57. For the 1909 Massachusetts commission ruling, see "Electric Rates-Massachusetts" (1912). L. R. Nash's text (1933, 321) gives other examples of regulatory commissions approving demand charges as instruments of price discrimination.

58. US Bureau of the Census 1905, 3.

59. Bushnell 1909.

60. Among these: "The Isolated Plant Versus the Central Station—Discussion at the Chicago Electrical Association" (1902); "Electrical Plant in the Newark Free Public Library" (1903); "Discussion of Papers on the Issue of Central Stations vs Isolated Plants" (1910); "Isolated Plants—Steam Heating" (1910); Bushnell 1909;

Carter 1909; Hale 1903; Hale 1910; Knowlton 1907; Main 1910; Maxwell 1909. Between 1908 and 1916, the periodical *Isolated Plant* was published in New York.

Chapter 3

1. In the 1990s, some states restructured the electric utility industry, dividing generation, transmission, and distribution into separate industries. A brief description of this is in the conclusion.

2. Details on the characteristics of various franchises given electric utilities can be found in Wilcox 1910, 2–3, 150–216; and National Civic Federation 1913, 1907b, 665–768.

3. Wilcox 1910, 209–10.

4. Ibid., 158, 163, 175.

5. Ibid., 143.

6. Troesken 2006, 269.

7. Examples include the franchise awarded Utah Power Company in Salt Lake City after the incident with Lacombe Electric Company (discussed later) and an interesting case given by Wilcox of a franchise given the New Haven Water Company in 1891. The rates charged by that company could be arbitrated every five years "provided, however, that the arbitrators should not fix the rates at a point too low to pay all operating costs, renewals, extensions, and every other kind of expense, including interest on funded and unfunded debt, and in addition thereto an 8 percent dividend on the company's existing capital stock, together with a reasonable return, not exceeding 8 percent, upon other capital that might be invested from time to time in additions or extensions to the plant." As Wilcox pointed out, this one-sided bargain enabled the company to earn a profit on all investment regardless of its contribution to the city. Wilcox 1910, 158, 163, 175. George Priest (1993, 321) argued that these arbiters may have operated like state commissions and that municipal regulation was evolving in the direction eventually taken by state regulation. There is little available historical evidence for this, and the legal requirements imposed by *Smyth v. Ames* would have required municipalities to devote considerable resources to the task, as was done by state commissions.

8. Wilcox 1910, 143.

9. In 1864, the territorial legislature of Colorado had granted the Occidental Gas Light Company an exclusive thirty-year franchise for gas lighting in Denver. Glaeser 1929, 201–2.

10. Wilcox 1910, 142–43.

11. Mills 1905, 488–92; Wilcox 1910, 142–43, 184–88.

12. Much of the material presented here appeared in Neufeld 2008. The term "rent" generally is applied to the price of a resource whose supply is not affected by price. This can be seen in the classic example of land: a higher rent for land does

not increase the total amount of land, nor does a lower rent decrease that amount. A monopolist is able to earn above-normal profits by withholding output to increase price. The higher price is associated with lower, not greater, output. The term "monopoly rents" refers to the resulting above-normal profits. Alchian 1987.

13. Quasi-rent is the difference between the value of capital if used for the purpose originally intended and its value if used in the next-best purpose. Transaction-specific capital has little or no value if not used as originally intended; its value equals quasi-rent. Payment for the capital equipment of an electric utility is rent-like because that equipment will continue to be used until it wears out even if its costs are not covered. Klein, Crawford, and Alchian 1978. The argument that transaction-specific capital investments led to corruption harmful to both utilities and their customers is found in Goldberg 1976; Troesken 2006.

14. If only monopoly profits were extorted, there should have been no long-run change in a utility's operations. Competition could clearly threaten quasi-rents, as happened in Denver. The threat of competition might have been used by honest officials in an attempt to lower rates. Such an action might have been politically popular, but if it reduced quasi-rents, it would not have been in the public's long-run interest.

15. Creamer, Dobrovolsky, and Borenstein 1960, 241–47; Ulmer 1960, 256–57, 320–21. Only privately owned electric utilities and factories with annual outputs of $500 or more are included.

16. Steffens 1904. See the chapters on the cities mentioned.

17. "$1,200 a Month Phone Graft for Boss Ruef" (1907, B1); Mowry 1951, 30–38.

18. McDonald 1962, 82–89; National Civic Federation 1907b, 676–77. Additional examples are found in Troesken 2006.

19. McCormick 1981.

20. Fisher 1907, 38.

21. Municipal ownership of waterworks as a form of vertical integration is argued by Troesken and Geddes 2003. Others have shown that public ownership in gas and telephone utilities was related to the inability of governments to make credible assurances to private companies that confiscation of quasi-rents would not occur. Levy and Spiller 1994; Troesken 1997.

22. The chart was developed by using a word search on the contents of the Pro-Quest historical newspapers database for entries whose contents concerned the issue of privately owned and government-owned public utilities. The search was limited to articles and editorials. The query was run for all dates between January 1 and December 31 for each year. Of course, this approach could not perfectly identify the target entries, but there is no reason to suspect bias over time. Proquest (2015).

23. Miller 1971, 26–32.

24. McLean 1900, 349–352; Miller 1971, 24–41.

25. Miller 1971, 26–30.

26. These commissions were established as "permanent," but they were occasionally abolished. White 1921, 178–84.

27. Cushman 1972, 22–26.

28. Miller 1971, 16–23.

29. Munn v. Illinois, 94 U.S. 113 (1877).

30. Wabash, St. Louis and Pacific Railway Company v. Illinois, 118 U.S. 557 (1886).

31. Chicago, Milwaukee and St. Paul Railway Company v. Minnesota, 1890. 134 U.S. 418 (1890); Barron 1942, 784–87; Smalley 1906, 45–48.

32. Smyth v. Ames, 169 U.S. 466 (1898).

33. The case involved a single class of freight carried entirely within the state, thus avoiding the question of interstate commerce. The court accepted the contention that rates for that service had to cover the costs of providing that service, and it accepted the method the railroad used for determining the portion of all operating costs assigned to that service. The railroads had no data on costs for the service in question. Instead, they had data only for freight that either originated or ended in Nebraska but may have come from or gone to another state. The railroads argued that the cost of transporting freight wholly within Nebraska was 10 percent higher than that for the freight for which they had data. Instead of adding 10 percent to determine the cost of intrastate state, the court added 10 percentage points, resulting in a 15.4 percent increase. It accepted the railroad's contention that a 29.5 percent reduction in maximum rates would result in the same percentage decline in revenue. This underestimated the actual revenue the railroads would have received because it ignored the fact that not all shippers were charged the maximum rate and that the lower rates would have resulted in more freight being shipped.

34. This view equates the setting of rates to the seizure of property under eminent domain. Other standards that form the basis of laws affecting business are not as limiting. A company's rate of return might be reduced by minimum wage laws or environmental or safety regulations, but a company does not have a Fourteenth (or Fifth) Amendment right to avoid the costs these impose.

35. Bauer 1926; Bauer and Gold 1939, 203–6, 232–43; Bemis 1927; Gruhl 1914; Hagenah 1932; Willis 1927.

36. Federal Power Commission v. Hope Natural Gas, 320 U.S. 591 (1944).

37. Interstate Commerce Commission v. Cincinnati, New Orleans and Texas Pacific Railway Company, 167 U.S. 479 (1897).

38. Amendment of an Act to Regulate Commerce (Hepburn Act), P. L. No. 59-337, 34 Stat. 584 (1906).

39. The discussion that follows is largely based on Smalley 1906, 83–128. Smalley appears to have been describing conditions contemporaneous with the paper's publication in 1906, prior to the resumption of rate regulation by the Interstate Commerce Commission.

40. Cushman 1972, 65–66.

41. An Act to Regulate Railroads and Other Common Carriers , 362 Wis. Stat. 549–69 (1905); Commons 1905, 69.

42. See, for example, "Appeal from Rates" (1906); "Are the Federal Courts Unworthy of Confidence?" (1906); "The Question of Review" (1906).

43. Maltbie 1912.

44. Wilcox v. Consolidated Gas Co., 19 U.S. 212 (1909).

45. Houston, East and West Texas Railway Company v. United States, 234 U.S. 342 (1914); McCollester 1922.

46. Insull 1898, 24–29. The spelling of "therefor" was common at that time.

47. McDonald 1958, 242–43.

48. "Report of Committee on Legislative Policy" (1904, 89). The committee produced no report, and it was then disbanded. Davis 1905, 7.

49. $10,000 in 1899 dollars ($282,000 in 2014 dollars) does not seem enough for a very extensive investigation.

50. Doherty 1902, 36–38.

51. "Control of Franchises" (1903); "The Municipal Ownership Convention" (1903); "Municipal Ownership Convention Meets" (1903).

52. "A New Civic Federation" (1899).

53. Wunderlin 1992.

54. Moffett 1907, 13.

55. A separate investigation of sixteen electric utilities in Massachusetts was designed to be less comprehensive than that conducted elsewhere in the United States. Those conducting the Massachusetts investigation were able to collect data from its private utilities. National Civic Federation 1907b, xiv.

56. National Civic Federation 1907a, 21–22.

57. Walton J. Clark, vice president of a Philadelphia gas company, vigorously, dissented, arguing that there was never a proper role for municipal ownership. National Civic Federation 1907a, 29–32. He joined Charles L. Johnson in a separate statement that asserted that if a municipality could take over a privately owned utility, the prior approval of elected representatives, in addition to a popular vote, should be required. Ibid., 28.

58. Ibid., 23.

59. For example, the *Wall Street Journal* in 1903 expressed the view that government regulation of privately owned utilities was the only alternative to government ownership. "The Question of Municipal Franchises" (1903).

60. The report did recommend consideration of the "sliding scale" rate structure, an early form of incentive regulation used by London's gas works and then in vogue. A utility that lowered its costs would divide the savings between lower prices and higher dividends to its stockholders according to a fixed formula. National Civic Federation 1907c, 146–47.

61. Jensen 1956, 237–38; McGuire and Granovetter 1998, 13–14. McGuire and

Granovetter, who examined detailed records of the NCF, argue that the investigation itself was thoroughly biased in favor of private ownership, a claim that had been earlier considered and rejected by Frank Parsons (1907), president of the National Public Ownership League, who served on the investigating committee.

62. Gray 1900; Jones 1979, 447–48.

63. "$45,500,000 of Water in New York Edison Co" (1905); "Electric Light Costs Small Consumer Dear" (1905).

64. "Gas Report Held up by Tammany Panic" (1905); "What May Happen" (1905).

65. Mosher 1962, 239.

66. The Public Service Commissions Law, 429 N.Y. U.C.C. Law 889 (1907).

67. Bruere 1908, 3.

68. An Act Giving the Wisconsin Railroad Commission Jurisdiction over Public Utilities, 499 Wis. Stat. 488–83 (1907).

69. Commons 1934, 111–28.

70. Knittel (2006) examined factors that would have been important to different interest groups and concluded that states in which residential users had the most to gain adopted regulation more quickly. He also found a similar effect in those states in which industrial users and coal interests stood to gain the most. Neufeld (2008) found that states with higher levels of per capita generating capacity, interpreted as indicating a higher level of quasi-rents, were quicker to adopt state regulation.

71. Hausman and Neufeld (2002).

72. The study by Knittel (2006) and that by Neufeld (2008) on the rate at which states adopted regulation imply that state regulation was expected to encourage investment. Knittel implicitly assumes that this was the expectation of the interest groups associated with quicker adoption. Neufeld found that states that more quickly adopted regulation were the ones whose utilities would have benefited the most from the protection of their investment state regulation offered. A reasonable inference from this would be that once a state was regulated, and future investment enjoyed greater protection, rates of investment would have increased. However, Lyon and Wilson (2012) found evidence that, despite the protections of capital it afforded, regulation had a negative effect on new investment in states that had rapidly growing populations.

73. Within the economics literature, the efficiency lost by this reduced management effort is termed "X-efficiency," and the seminal article is Leibenstein (1966). A huge literature on the topic within economics generally confirmed that regulation causes a loss of this type of efficiency.

74. The overuse of capital equipment resulting from rate-of-return regulation was termed the Averch-Johnson or A-J effect after the authors who first proposed it. Averch and Johnson 1962. Most of the resulting work occurred in the 1960s and early 1970s.

75. Baumol and Klevorick 1970.

76. Competition and the fact that many products can be stored or resold tend to result in similar price structures across firms in the same industry based on differences in the cost of serving different customers.

77. The seminal work making this argument is Peltzman 1976. Peltzman's theoretical argument assumes different consumer groups desire lower prices, even at the expense of the prices charged other groups. Meanwhile, the utility desires higher profits. Peltzman assumed that the political influence of each of these groups was subject to diminishing marginal returns, and this would encourage regulatory agencies to avoid giving one group everything it wanted at the expense of another group. A consequence would be the tendency to reduce the differential between the rates charged different groups.

78. Morrison and Winston 2000, 1.

79. US Bureau of the Census 1930, 66, 68.

80. "Discussion on Interconnection and Superpower" 1924, 219–220.

81. Public Utilities Commission of Rhode Island v. Attleboro Steam & Electric Company, 273 U.S. 83 (1927).

Chapter 4

1. US Energy Information Administration 2014a.

2. Carter et al. 2006, tables Db 234–37.

3. Schroeder 1986. The illustrations are from Keene 1918, 327, 379. Other appliance sockets were shown but these did not include the modern wall socket.

4. Platt 1991; see reproduction of GE advertisement between pages 110 and 111.

5. Ibid., 241.

6. Du Boff 1964, 62, 184–206. Du Boff corrected the more commonly used census data. US Bureau of the Census 1966, 6–9.

7. The 1917 special census categorized electricity as used for "light" or "power." Some electricity was also used for heat, but the census included it with power. The fact that heat was not given a separate category, and that the category in which it was included was called power, suggests that it was a small portion of the total. US Bureau of the Census 1920, 82.

8. Carter et al. 2006, tables Dd 852–53.

9. The relative share of residential use grew after this, but electricity sales to industrial users remained over twice that of sales to residential users until after the 1950s. Cintrón and Edison Electric Institute 1995, 243–44, 399–400. This does not itself imply that residential users were overcharged and industrial users undercharged. See the discussion of price discrimination in chapter 3.

10. Devine 1983; Du Boff 1964.

11. Scott 1897, 8–9.

12. Hunter and Bryant 1991, 115–84.

13. The average size of electric motors used in factories declined from 18.21 horsepower in 1899 to 10.99 in 1919 to 8.45 in 1939. US Bureau of the Census 1940, 275.

14. The growth rates in total factor and labor productivity for 1929–1948 were 1.7 and 1.6 percent, respectively. Growth in capital productivity showed an even more marked change in 1919. Between 1869 and 1919, capital productivity fell at an average compounded rate of 1.8 percent. Between 1919 and 1929 and between 1929 and 1948 it grew at rates of 4.2 and 2.2 percent, respectively. Kendrick 1961, 464.

15. A separate value for only transmission belts was not given for earlier years. Belts for clothing were produced by a different industry.

16. Gray 2013.

17. Baruch 1941, 38–39, 298–99.

18. *Emergency Power Bill Hearings* (1918, 19).

19. Insull 1924, 150–151.

20. Murray 1925, 46.

21. Murray et al. 1921.

22. Ibid., 12.

23. Ibid., opposite 141.

24. Ibid., 12, 147

25. Ibid., 14.

26. "Report of the Superpower Survey Committee" (1921).

27. Much of the material on the attempt to implement a superpower system is drawn from DeGraaf 1990.

28. Murray 1925, 154.

29. Hughes 1983, 308.

30. *Report of the Giant Power Survey Board to the General Assembly of the Commonwealth of Pennsylvania* (1925).

31. Christie 1972, 487–88.

32. Christie 1972; McGeary 1960.

33. Pinchot 1925, iii-xiii.

34. *Report of the Giant Power Survey Board to the General Assembly of the Commonwealth of Pennsylvania* (1925, 1–2).

35. Distributors could own and operate relatively small (less than 25,000 kW) generators.

Chapter 5

1. Field 1932.

2. More than one-third of the stock was owned by the Duke Endowment, which could be sold only under restrictive conditions. Another 27 percent was owned

by a trust for the benefit of J. B. Duke's daughter. US Federal Trade Commission 1935c, 89–90, 109–11; US House Committee on Interstate and Foreign Commerce 1934b, 431–432.

3. US House Committee on Interstate and Foreign Commerce 1934b, 264–322.

4. Details on the variations that existed within each class of security at that time is provided in US Federal Trade Commission 1935a, 371–405.

5. Unless otherwise noted, information on Electric Bond & Share comes from Bonbright and Means 1932, 98–108; US Federal Trade Commission 1927, 69–7; US Federal Trade Commission 1928a, 99–100; Mitchell 1960; US Federal Trade Commission 1935a, 86–92.

6. US Federal Trade Commission 1927, 3–4, 69–78; US Federal Trade Commission 1935a, 86–92.

7. US Federal Trade Commission 1928a, 99–100.

8. Bonbright and Means 1932, 107–8; US Federal Trade Commission 1935a, map opposite p. 88.

9. Waterman 1936, 99–124. An investor could have much more effectively enjoyed the benefits of diversification by investing in different industries.

10. Figures 5.6 and 5.7 are modified versions of maps produced by the Federal Trade Commission that showed counties served by operating companies controlled by holding companies. These are shown in both figures as light gray. The original maps provided information on many holding companies and subholding companies, but inconsistent patterns were used for the territories of the same holding companies in the two years. To improve clarity, the original maps have been digitally enhanced and overlays highlighting the territories of Electric Bond & Share, the Insull group, and (for 1929) the United Corporation have been applied.

11. Unless otherwise noted, material on Stone & Webster comes from "Stone & Webster" (1930); Hughes 1983, 386–93; US Federal Trade Commission 1927, 167–87; US Federal Trade Commission 1935a, 96, 643–47.

12. US Federal Trade Commission 1928b, 210.

13. Unless otherwise noted, the information about the United Corporation comes from Bonbright and Means 1932, 127–138; U.S. Federal Trade Commission 1935a, 111–16.

14. US Federal Trade Commission 1928a, 216–18.

15. Bonbright and Means 1932, 133; Buchanan 1936, 48.

16. McDonald 1962, 250.

17. Prepared by the author based on a similar diagram in US Federal Trade Commission 1935a, opposite 114. Some of the holdings represented by arrows were small.

18. Taft and Heys 2011, 75–81; US Federal Trade Commission 1935a.

19. Bonbright and Means 1932, 127–138.

20. Unless otherwise noted, information in this section comes from Hughes 1983; Insull and Plachno 1992; McDonald 1962; Platt 1986; US Federal Trade Commission 1935a.

21. An interesting example of his risk taking occurred with his adoption of the turbine generator, the first in the United States. During its testing, he insisted on putting himself at physical risk by being present against the advice of the engineer in charge of the project. McDonald 1962, 99–100.

22. Ibid., 277.

23. McDonald forcefully argued (ibid.) that, rather than being a predatory crook, Insull was a man with high moral and ethical principles who was unfairly made a scapegoat. Thomas Hughes (1983, 103–4) agrees that Insull was treated unfairly and that he did not deserve the opprobrium he received.

24. Wigmore 1985, 345.

25. Prepared by the author based on US Federal Trade Commission 1935a, opposite 160. Omitted from the figure but shown in the original report were the security types of all financing for the subholding and operating companies. Georgia Power & Light was a combination operating and holding company. The FTC also showed information on other operating companies controlled by Seaboard Public Service Co.

26. US Senate Committee on Banking and Currency 1934, 384.

27. Taylor 1962.

28. Sullivan 2010, 59; Taylor 1962, 188–89.

29. "Insull Rose to Top of Utilities Empire" (1938); "Samuel Insull" (1938).

30. "Text of Governor Roosevelt's Speech at Commonwealth Club, San Francisco" (1932).

31. "Samuel Insull" (1938).

32. Buchanan 1936, 44.

33. Ibid., 45–47.

34. Both tables 5.1 and 5.2 omit Cowles indexes, such as "utilities" that combined other indexes. Cowles "C" indexes, which assumed that dividends were reinvested in the same industry, were used. The index values were adjusted for changes in the price level by the Consumer Price Index. Table 5.1 includes all indexes with data in January 1920. Table 5.2 includes all indexes with data in January 1928.

35. These are simple percentage changes from the previous year using index values uncorrected by the CPI.

36. "Alex Dow Denounces Speculation" (1925).

37. McGrattan and Prescott 2004; White 1990.

38. "Fisher Says Prices of Stocks Are Low" (1929); "Says Stock Slump Is Only Temporary" (1929).

39. Excluding indexes that combined other indexes, the total number was 62. Cowles (1939, 4015–18) provided earnings/price ratios in his series R.

40. McGrattan and Prescott 2004, 1006–8.

41. US Bureau of Corporations 1912, 180.

42. US Department of Agriculture 1916, 3.

43. Buchanan 1936, 43–44.

44. Ibid., 46–47.

45. Quoted in Lowitt 1971, 359.

46. Quoted in Funigiello 1973, 4.

47. US Federal Trade Commission 1927, xvii.

48. Ibid.

49. Ibid., 50.

50. Ibid., 36–37.

51. Ibid., v, 49–50, 172–75.

52. Davis 1962.

53. For example, in the earlier report, Pacific Gas & Electric (PG&E) and Montana Power were listed as holding companies. The 1935 report categorized both of these as large local companies, although very large blocks of their stock were held by a holding company, 29.7 and 28.5 percent, respectively. It is unclear why these companies were not placed in holding company systems. US Federal Trade Commission 1927, 36, 266–67; US Federal Trade Commission 1935a, 37–39.

54. Ulmer 1960, 476–477. This refers to the change from 1929 to 1932.

55. Carter et al. 2006, tables Db 234–41.

56. US Securities and Exchange Commission 1959, xvii-xviii.

57. Cowles 1939, 168–266.

58. US Securities and Exchange Commission 1959, xvii-xviii.

Chapter 6

1. A recent illustration of the difficulties of forming an integrated network of utilities under separate ownership was provided by Wald 2013.

2. The holding companies currently operating those networks are The Southern Company, Entergy, and American Electric Power, respectively.

3. *Public Utility Holding Companies. Part 3* (1935, 2294); US Federal Trade Commission 1935a, 43.

4. US Securities and Exchange Commission 1952, 110–11.

5. US Federal Trade Commission 1935b, 63.

6. Over 100 articles mentioning the FTC hearings on holding companies were published in the *New York Times* between 1930 and 1935. Additional examples of articles in other newspapers can be found in McDonald 1962, 268. Books written for a popular readership include Gruening 1964 [1931]; Levin 1931; Ramsay 1937; Raushenbush 1928; Raushenbush 1931; Thompson 1932.

7. Pecora 1939; US Senate Committee on Banking and Currency 1934.

8. Securities Act of 1933, P. L. No. 73-22, 48 Stat. 74 (1933); Securities Exchange Act of 1934, P. L. No. 73-291, 43 Stat. 881 (1934).

9. Dillavou 1933; Mosher 1929.

10. US Federal Trade Commission 1935a, 8–16.

11. One exception may be work done by Forrest McDonald in the research for his biography of Insull. That research included the extremely critical evaluation by the FTC of the companies in the Insull group. He also reviewed other sources,

including the transcripts of Insull's trial. McDonald repeatedly accused the FTC reports of being biased and unreliable, but he provided no basis for his accusations. McDonald did acknowledge that some actions taken by Insull's companies once collapse seemed imminent were questionable. McDonald 1962, 126–27, 275, 292, 295, 298.

12. US Federal Trade Commission 1935a, 512–18.

13. Ibid., 12–62; US Federal Trade Commission 1935b, 11–12.

14. Buchanan 1935; US Federal Trade Commission 1930, 935, 942–43, 953; US Federal Trade Commission 1935a, 235–36, 257–58.

15. U.S. Federal Trade Commission 1935a, 845.

16. Healy 1936, 314–18; US Federal Trade Commission 1935a, 812–24.

17. One exception was Duke Power. US Federal Trade Commission 1935a, 509–12.

18. Dorau 1930, 618–21; Young 1965, 33.

19. US Federal Trade Commission 1935a, 502–11.

20. US Federal Trade Commission 1934, 243.

21. Jones and Bigham 1931, 486.

22. Mosher and Crawford 1933, 149–51.

23. US Federal Trade Commission 1935a, 565–83. Insull maintained the actions were taken to block a hostile takeover. Insull and Plachno 1992, 188–202.

24. US Federal Trade Commission 1935a, 535. US Senate Committee on Banking and Currency 1934, 93–101.

25. This behavior received a great deal of notice during the savings and loan crisis of the 1980s when it was encouraged by the moral hazard created by deposit insurance. It also, however, was seen in the 1930s during the crisis thrift institutions then encountered. Grossman 1992.

26. US Federal Trade Commission 1935a, 238–39, 445–47.

27. US House Committee on Interstate and Foreign Commerce 1934a, 76.

28. US Federal Trade Commission 1935a, 14.

29. Ibid., 358–59, 877.

30. Ibid., 160–61.

31. Concern over the divergence of interests that might exist between stockholders and management and the implications of this for the way companies were run arose after World War II. The most influential early work was Berle and Means 1968. Fifteen years later the *Journal of Law and Economics* devoted a special issue to this topic and their work (vol. 26, no. 2, June 1983)

32. Bonbright and Means 1932, 143–44.

33. US Federal Trade Commission 1935b, 28. The regulations referred to here were not limited to those administered by state utility regulatory commissions but include regulations concerning securities and the granting of corporate charters.

34. US Federal Trade Commission 1934, 120–22.

35. Ibid., 260.

36. US Federal Trade Commission 1935b, 6–7.

37. US Federal Trade Commission 1935a, 851–52. Unlike electricity, the FTC found that wholesale gas prices had been manipulated.

38. US Federal Trade Commission 1935a, 18–19, 215, 462–68, 599–670, 842–45; Bonbright and Means 1932, 186–87.

39. Bemis 1927; Heilman 1914; US Federal Trade Commission 1935b, 6–7.

40. Healy 1936, 318; US Federal Trade Commission 1935a, 615–65.

41. US Federal Trade Commission 1934, 312; US Federal Trade Commission 1935b, 16.

42. Bonbright and Means 1932, 164–66; US Federal Trade Commission 1935b, 16–18.

43. U.S. Federal Trade Commission 1935a, 238–39; US Federal Trade Commission 1935b, 13–16.

44. Bonbright and Means 1932, 163–75; Buchanan 1935; Healy 1936, 298–99; US Federal Trade Commission 1935b, 16–18.

45. Bonbright and Means 1932, 167–74; US Federal Trade Commission 1935b, 17, 358–60.

46. US Federal Trade Commission 1935a, 657–58.

47. US Federal Trade Commission 1934.

48. McDonald 1962, 165–87.

49. Levin 1931, 53–55.

50. US Federal Trade Commission 1934, 15–16.

51. Ibid., 70–77.

52. Ibid., 235.

53. *Public Utility Holding Company Act of 1935* (1935, 355); Heilman 1925; US Federal Trade Commission 1934, 11–17, 302–7; US Federal Trade Commission 1935a, 348–49.

54. US Federal Trade Commission 1934, 61–110.

55. Ibid., 88–91.

56. Ibid., 186–212.

57. Ibid., 111–38.

58. Ibid., 133–34, 319–90.

59. Ibid., 308–13.

60. Ibid., 131, 314–17.

61. Armstrong and Nelles 1986; Evans 1992; Fleming 1992; Freeman 1996; Lowitt 1968.

62. Lowitt 1968; US Federal Trade Commission 1934, 356–362.

63. In 1945, the industry created a new organization, the National Association of Electric Companies (NAEC), headquartered in Washington to resume lobbying activities. In 1979, the EEI moved its headquarters to Washington and absorbed the NAEC. Crickmer 1993.

64. Barnard 1966.

65. Roosevelt and Rosenman 1938, 78.

66. Bellush 1955, 208–42; Davis 1985, 91–92.

67. Roosevelt and Rosenman 1938, 727–42. For the history of the formation of the yardstick role by Roosevelt, see McCraw 1971, 30–31.

68. Funigiello 1973, 43.

69. "Geographic Integration under the Public Utility Holding Company Act" (1941).

70. *Public Utility Holding Company Act of 1935* (1935, 223–86). Similar testimony was given before the House committee.

71. "The Lobby Inquiry: Mr. Hopson Loses at Hide and Seek" (1935); "Protests against Rayburn-Wheeler Bill Rise to Extraordinary Proportions" (1935); Funigiello 1973, 98–121.

72. *Public Utility Holding Companies. Part 3* (1935, 1653).

73. Ibid., 1612.

74. *Investigation of Public Utility Corporations, Part 2* (1928, 35–36).

75. *Public Utility Holding Companies. Part 3* (1935, 2294).

76. Plum 1938, 150.

77. US Federal Trade Commission 1935b, 1–28.

78. Marlett and Marple 1935; Marlett and Traylor 1935c; Marlett and Traylor 1935b.

Chapter 7

1. Wilson 2002.

2. Swain 1888.

3. There were smaller hydroelectric plants at Niagara Falls as early as 1893. Belfield 1976; Hunter and Bryant 1991, 254–72; Prins 1991.

4. "The Great Southern Transmission Network" (1914).

5. "Report of the National Conservation Commission" (1909, 6).

6. Rivers and Harbors Act, 1925, P. L. No. 68-586, 43 Stat (1925).

7. Rivers and Harbors Act, 1927, P. L. No. 69-560, 44 Stat (1927); US Army Chief of Engineers and Secretary of the US Federal Power Commission 1926.

8. River and Harbor Act, 26 Stat. 426 (1890); Billington, Jackson, and Melosi 2005, 121–22.

9. River and Harbor Act, 30 Stat. 1121 (1899).

10. Cushman 1972, 275–78; Kerwin 1926, 114–263.

11. Federal Water Power Act, P. L. No. 66-280, 41 Stat. 1063 (1920).

12. Cushman 1972, 280–83; Kerwin 1926, 321–22.

13. An Act to Reorganize the Federal Power Commission, 46 Stat. 797 (1930); Cushman 1972, 280–97.

14. Pisani 1992, 76.

15. The federal government did not displace other providers of irrigation. The federal share of total irrigated acreage in seventeen western states never exceeded 15 percent until after 1930 and remained below 25 percent thereafter. Wahl 1989, 17.

16. Pisani (1992) has provided a history of western irrigation through the passage of the Reclamation Act.

17. References to the bureau also include the service when referring to pre-1907 events.

18. Burness, et al. 1980; Gressley 1968, 251–54.

19. Pisani 2002, 214.

20. Town Sites and Power Development Act, P. L. No. 59-103, 34 Stat. 116 (1906).

21. Pisani 2002, 214–20. At one project, a 148-mile transmission line was constructed.

22. James 1917, 65–85; US National Park Service 2012.

23. Kleinsorge 1941, 46–47.

24. The total quantity carried by the river each year was estimated sufficient to cover 214 square miles one foot deep. Kleinsorge 1941, 12.

25. Homan 1931, 181–82; Stevens 1988.

26. Hughes 1986, 388–89; Kleinsorge 1941, 21–54; Stevens 1988, 12–19.

27. Kahl 1983. Gary Libecap has written on this issue and the property rights problems involved. Libecap 2005; Libecap 2007.

28. Kleinsorge 1941, 52.

29. Ibid., 186–90.

30. Fall and Davis 1922, 1–21.

31. Colorado River Compact Act, P. L. No. 67-56, 41 Stat. 171 (1921).

32. "Colorado River Compact" (1922).

33. Arizona had a couple of other issues as well, including the right to tax the electricity produced by the Boulder Canyon Dam. Kleinsorge 1941, 66–71.

34. Boulder Canyon Project Act, P. L. No. 70-642, 45 Stat. 1057 (1928).

35. The act also authorized the construction of the All-American Canal, not paid for by electricity sales from Hoover Dam.

36. Leatherwood 1928, 136.

37. Kleinsorge 1941, 182–83.

38. Moeller 1971, 26.

39. Greenwood 1928, 84–104; Homan 1931, 212–15.

40. US Federal Trade Commission 1934, 331–46.

41. Leatherwood 1928; Moeller 1971, 30.

42. Wilbur and Ely 1933.

43. US Bureau of the Census 1934, 36.

44. Wengert 1952.

45. Nitrate Division 1922, 1–11.

46. National Defense Act, P. L. No. 64-85, 39 Stat. 166 (1916).

47. An excellent account of the entire history of Wilson Dam can be found in Hubbard's (1961) work, from which much of the following discussion is drawn.

48. Schlesinger 1959, 322.

49. Congress approved completion of Wilson Dam in 1923 before resolving the Ford bid.

50. Hubbard 1961, 28–47.

51. Ibid., 48–146.

52. Neufeld, Hausman, and Rapoport 1994.

53. Hoover 1931; Hubbard 1961, 147–315.

54. Tennessee Valley Authority Act, P. L. No. 73-17, 48 Stat. 58 (1933).

55. Commonwealth and Southern Corporation 1937, 51–52.

56. Tennessee Valley Authority 1935, 23.

57. A 1959 act passed by Congress expanded the TVA's borrowing authority, replacing contentious congressional appropriations that had previously been necessary. It also prevented the TVA from expanding its service area and displacing private utilities. This eliminated a role for TVA as a "birch rod." Amendments to TVA Act P. L. No. 86-137, 73 Stat. 280 (1959).

58. Quoted in Barnard 1966, 89.

59. Barnard 1966, 85; Lilienthal 1964, 197; McCraw 1971, 42.

60. "Wendell Lewis Willkie" (1937, 195).

61. Roosevelt 1933.

62. An analysis of properties acquired by the TVA around Guntersville Dam between 1936 and 1939 found that the courts sometimes did lower the valuation given by the TVA, and often came up with the same valuation. On average, however, owners that went to court were awarded 5 percent more than the TVA offer. Kitchens 2013b.

63. In a 1935 amendment to the TVA Act, excess revenue could be used to cover the manufacturing costs of fertilizer and dam operation. The language of the act is permissive rather than obligatory. Any revenue above that was to be paid to the US Treasury. Amendments to TVA Act, P. L. No. 74-412, 49 Stat. 1075 (1935).

64. US House Committee on Rivers and Harbors 1930, 88.

65. Rather than nitrates, which had been the central purpose of Wilson Dam, the TVA emphasized the use of phosphate fertilizer.

66. McCraw 1971, 88–89.

67. Talbert 1987, 36–37.

68. McCraw 1971; Talbert 1987.

69. Billington, Jackson, and Melosi 2005, 358–62; US House Committee on Rivers and Harbors 1930.

70. Talbert 1987, 111–16.

71. For conditions at Hoover Dam see Stevens 1988, 52–55, 62–71, 117–40.

72. McCraw 1971, 30–31.

73. Ibid., 53–59.

74. Amendments to TVA Act, P. L. No. 74-412, 49 Stat. 1075 (1935).

75. The original TVA Act required payment of 5 percent of its gross revenue. In a 1940 revision of the act, the proportion was increased and a minimum payment rule was established that determined the payments actually made. The TVA was directed to take the average of the taxes collected on the properties two years before their acquisition. The TVA then paid 40 percent of that average, the percent of their value allocated to power production. Municipalities were allowed by their contracts with the TVA to take in lieu of taxes an amount equal to the sum of the state, county, and municipal rates multiplied by the book value of the electrical systems they owned, amounts similar to that which a privately owned company would have paid. Emergency Relief Appropriation Act, P. L. No. 76-88, 54 Stat. 611 (1940); Harbeson 1936; Pritchett 1942; Tennessee Valley Authority 1941, 22–24; Tennessee Valley Authority Act, P.L. No. 73-17, 48 Stat. 58 (1933).

76. "Utility Men Hail Birmingham Vote" (1933); Bonbright 1940, 47.

77. McCraw 1970.

78. McCraw 1971, 109.

79. Ashwander v. Tennessee Valley Authority 297 U.S. 288 (1936).

80. Tennessee Electric Power Co. et. al. v. Tennessee Valley Authority et. al., 306 U.S. 118 (1939).

81. McCraw 1971, 108–21; Purcell 2002.

82. Carter et al. 2006, tables Db 234–41; Field 1990, 35; Roosevelt 1933.

83. Cannon 1934.

84. Atkins 2006, 190–91.

85. By the beginning of 1940, the immediate rate had been eliminated for almost all communities served by Alabama Power. Tennessee Electric Power had been sold to the TVA and no longer operated. Georgia Power, however, continued to make extensive use of immediate rates. US Federal Power Commission 1940.

86. Newton 1935.

87. The TVA sold directly to some large industrial customers. It also would temporarily function as a distributor until a permanent distributor could be organized. There were two cases in which the TVA sold power to a small local distributor that was privately owned. The model of a government-owned producer of electricity only supplying electricity at wholesale and only to municipally owned utilities followed that used by Ontario Hydro.

88. Chapters 2 and 3 contain discussions of the origin and effect of demand-charge rate schedules that used maximum power usage to determine rates.

89. Copies of the contracts with distributors were published in all of the TVA's annual reports except the first.

90. Tennessee Valley Authority 1956, 8–9.

91. These are average state revenues from the 1934 census report. The census data combine Alabama and Mississippi and do not enable determining the separate average rates for each state. This was done with some other states as well. C&S had not yet introduced its new lower rates. US Bureau of the Census 1934, 66–67.

92. "Transactions on the New York Stock Exchange" (1933, 20); "Utility Issues Hit by Federal Rates" (1933, 19). This drop in stock values is indicative of stockholder concern, but it is not conclusive since many other factors would affect stock prices on a single day.

93. The TVA Act (1933, Section 14) required the authority to determine the value of Wilson Dam and to allocate its cost among the different functions, including fertilizer and national defense, but the authority decided not to include those two functions in the cost allocation.

94. US Joint Committee on the Investigation of the Tennessee Valley Authority 1939, 284.

95. Porter 1938a.

96. Morgan 1938, 5.

97. *Investigation of the Tennessee Valley Authority, Part 10* (1938, 4222–23). An interesting analysis of the TVA's allocation problem used game theory in which each of the different uses of the dam was treated as a player in a cooperative game to devise a feasible division of the "nonseparable" joint costs. One of the factors in this analysis (and in the analysis undertaken by the TVA) was the cost of a dam designed solely to perform a single purpose. The total cost of a multipurpose dam is, of course, significantly less than the sum of the costs of single purpose dams, but those costs enter into a determination of the feasibility of any division of the non-separable cost of a multipurpose dam. The cost allocation used by the TVA was a feasible solution, and it was close to that adopted in 1950 as the recommended method for all water projects and used in water resources texts. It is, however, only one of an infinite set of feasible allocations. Straffin and Heaney 1981.

98. US Federal Power Commission 1940. The table shows average bills for January 1940 weighted by the populations of the towns. Average bills were only provided by for towns with a population of at least 250; rural areas were not included. The FPC gathered these rates by asking the distributors in each town to provide the amount charged for different levels of usage. When a utility had both immediate and objective rates, the FPC reported the size of bills calculated under each rate with no indication of how much electricity was sold under each. Only the lower objective rates were used in table 7.1. This almost surely understates the actual average bills paid in the non-TVA areas where both rates were used. A spot check of differences in the bills charged by different TVA distributors showed that they were consistent with the use of surcharges.

99. *Tennessee Valley Authority, Vol. 2* (1935, 589–90); Coppock 1940; Field 1990.

100. Tennessee Valley Authority 1939, 12.

101. US Bureau of the Census 1939, 62–63, 70–71.

102. Ottawa's rates were for 1937. Joint Committee on the Investigation of the Tennessee Valley Authority 1939, 214.

103. A typical case was Knoxville, where the vote was almost two to one in favor of creating a municipal agency to distribute TVA power. In Memphis, the vote was

seventeen to one. Birmingham, however, rejected TVA power. "Knoxville Votes Municipal Power" (1933); "Utility Men Hail Birmingham Vote" (1933); Hon 1940.

104. Kruesi and Vennard 1936.

105. Tennessee Valley Authority 1935, 261–62; "PWA-TVA 'Coercion' Charged by Utility" (1936); *Investigation of the Tennessee Valley Authority, Part 10* (1938, 4244–46, 4266); McCraw 1971, 124–30.

106. Tennessee Valley Authority 1935, 23. When the city of Knoxville was unable to come to terms with the Tennessee Public Service Company, a subsidiary of Electric Bond & Share, it began construction of a new system. The two then came to a deal with Knoxville paying more for the existing system than a new one would have cost. Bonbright 1940, 45; Tennessee Valley Authority 1940, 57–58.

107. "PWA-TVA 'Coercion' Charged by Utility" (1936). In March 1935, Chattanooga voted two to one in favor of a municipal utility. Through the use of legal challenges, the private utility serving the city did not relinquish its property until four years later when the TVA purchased Tennessee Electric Power from C&S. "Chattanooga Vote Is for T.V.A. Power" (1935).

108. *S. 1796 to Amend the Tennessee Valley Authority Act of 1933* (1939, 98–100); McCraw 1971, 152–55.

109. "Texts of Roosevelt Speeches Urging Public Operation of Power Plants" (1934); Commonwealth and Southern Corporation 1937, 51–52.

110. *Conservation and Development of the National Resources* (1937, 1–20); Roosevelt 1937.

111. McCraw 1971, 150.

112. Barnard 1966, 88.

113. The Central Electricity Board had been set up to improve the efficiency of Britain's electric utility industry, which had stagnated. The Central Electricity Board was created as a government-owned corporation with safeguards designed to insulate it from political pressure. It controlled a transmission system interconnecting most of the country's utilities. The most efficient generating plants were put under the board's control and were dispatched in the most efficient way. Their owners received payment guaranteed sufficient to cover all costs. Wholesale electricity was sold to local utilities for distribution. Chazeau 1934a; Chazeau 1934b; Hannah 1979, 105–49; Hormell 1932.

114. "Will Confer with Utilities" (1936); Ely 1936; Krock 1936; McCraw 1971, 91–107; Terral 1936.

115. Lilienthal 1964, 654–55; McCraw 1971, 143–48.

116. Being adjacent to the Tennessee Valley has high statistical significance in explaining the percentage change in residential rates but is not significant in explaining the change in all rates or in nonresidential rates.

117. Kitchens 2013a; Tennessee Valley Authority 1940, 7–11.

118. Kitchens 2014; Kline and Moretti 2014.

119. Carter et al. 2006, table Ca 11.

120. Fishback, Horrace, and Kantor 2005.

121. Leuchtenburg 1952; Owen 1973, 234–37.

122. "Columbia River and Minor Tributaries, Vol I" (1933); "Columbia River and Minor Tributaries, Vol II" (1933).

123. A discussion of the various proposals and bills considered by Congress can be found in Voeltz 1962. Further discussion of the politics surrounding the debate is in Funigiello 1973, 174–225; Ogden 1949.

124. Despite their complaints, farms in Washington and Oregon were much more likely to have utility service than farms in the rest of the country. In 1930, 41 percent of Washington farms and 27 percent of Oregon farms had utility service, while only 10 percent of all farms in the nation had such service.

125. Variations on this form of utility organization already existed in four states, and it had been advocated by the Giant Power proposal. Twentieth Century Fund 1948, 389.

126. Ogden (1949, 43–100) provides a discussion of the history of government versus private ownership of utilities in the Pacific Northwest and the development of PUDs.

127. Voeltz 1962, 73.

128. Magnusson 1937.

129. As of 2001, the river system had twenty-nine major dams controlled by the Corps of Engineers or the Bureau of Reclamation and dozens of privately owned dams. A complex set of agreements and a treaty with Canada were designed to enable the Columbia River system to operate as if it were under unitary control. A set of committees and working groups jointly planned for the operation of the system for periods ranging from one month to six years. Trade-offs between power and other needs were under the control of the individual dam operators. By 2001 environmental concerns, particularly the protection of migratory fish, had become a major consideration in the operation of the dams. The BPA dispatched power from ten of the dams, which gave it minute-to-minute control over their generation, but a different agency, the Grant County PUD, dispatched power from dams owned by a number of different utilities, and the power production of some of the dams were controlled by their owners. The "Hourly Coordination Agreement" kept the separate authorities in communication, enabling them to perform their dispatch responsibilities in concert. The BPA also controlled a single nuclear plant. Apart from that plant, the real-time operation of the system did not include thermal generation as in the TVA system. Federal Columbia River Power System 2001.

130. The FPC's allocation of the joint costs of Bonneville was 32.5 percent to power, less than the TVA's proportion, and the remainder to navigation. In the first year, however, only 6.5 percent of joint costs were allocated to power (Bonneville Power Administration 1939, 69–70).

131. Twentieth Century Fund 1948, 527–28.

132. As with the TVA contracts, if the rates were too low to permit a distributor to cover its costs, a mechanism existed to enable higher rates. Bonneville Power

Administration 1940a, 88–170; Twentieth Century Fund 1948, 519. In 1940 there were four government-owned utilities selling BPA power. For the first 50 kWh used per month, two charged 3¢ and two 4½¢ per kWh. All charged 2¢ per kWh for the next 50, 1¢ for the next 200, ½¢ for the next 900, and ¾¢ per kWh for all usage above 1200 kWh. These amounted to reductions of from 10 to 31 percent below previous rates. Bonneville Power Administration 1940b, 39–40. The BPA later discontinued efforts to control retail rates.

133. Bonneville Power Administration 1940b, 28–29.

134. These results undoubtedly helped doom proposals to give the BPA the power to acquire generating plants and create TVA-like geographic service areas. *Bills to Amend the Bonneville Act* (1942, 559–560).

135. Ogden 1949, 307–86.

136. Twentieth Century Fund 1948, 533.

137. Bonneville Power Administration 1940b, 42–45.

138. Cintrón and Edison Electric Institute 1995, 75–55; US Energy Information Administration 2014b.

Chapter 8

1. Carter et al. 2006, table Da 16; US Bureau of the Census 1975, part 2, 827.

2. US Bureau of the Census 1932, 540.

3. Brown 1980, xiv.

4. Middle West Utilities Company 1930, 101.

5. US Rural Electrification Administration 1938b, 67.

6. Caro 1982, 505.

7. Ibid., 511. Caro devotes an entire chapter (27) to the terrible drudgery experienced by women in farms but not in cities.

8. Brown 1980, xv-xvi; Morgan 1936. Arthur Morgan, then a director of the TVA, also advocated rural electrification to encourage the development of industry in rural areas.

9. Carter, et al. 2006, table Da 1295.

10. *Report of the Giant Power Survey Board to the General Assembly of the Commonwealth of Pennsylvania* (1925, 256–58).

11. Cannon 2000, 141; Cooke 1925, 55; Reed 1936, 756; Slattery and Mittell 1940, 5. It was often unclear whether the quoted prices covered just the lines or also other necessary equipment such as transformers. The terrain of a specific line could have a large impact on the cost.

12. Carter, et al. 2006, Da 1295, adjusted by the BLS Consumer Price Index.

13. Brown 1970, 38; Caro 1982, 516–17; Wolfe 2000, 524.

14. Reed 1936.

15. Adams 1936.

16. Middle West Utilities Company 1930, 1.

17. *What About Rural Electrification? Proceedings of the Farm Electrical Conference* (1926, 29).

18. US Bureau of the Census 1932, 538.

19. Spence 1962.

20. Learned et al. 1911.

21. Ibid. 1911.

22. Walthall et al. 1912.

23. Parker 1913.

24. Martin 1914, 146–49; Martin 1915, 152–60; Martin 1916, 118–27.

25. Martin 1915; Martin 1916.

26. "Report of the Rural Lines Committee" (1922).

27. Reed 1936, 743; Wolfe 2000, 520; Zinder 1929, 81.

28. Contemporary economists understood this. Zinder 1929.

29. A point made by some in the industry. Parker 1913, 139.

30. *Report of the Giant Power Survey Board to the General Assembly of the Commonwealth of Pennsylvania* (1925, 299).

31. "Report of the Rural Lines Committee" (1922, 106); Reed 1936, 750–752; Van Derzee 1928, 1109.

32. Brown 1980, 4–12; McCrory 1930; National Electric Light Association 1932, 5–13; Ronayne 1946, 6–9.

33. Walker 1928, 36.

34. Brown 1970; Brown 1980, 6–7; Ronayne 1946, 6–7.

35. Brown 1970, 23–34; Brown 1980.

36. *Report of the Giant Power Survey Board to the General Assembly of the Commonwealth of Pennsylvania* (1925, 37–40, 117–140, 243–305, 411–480).

37. Wells 1927, 14.

38. Two million people in the state, 40 percent of whom lived on farms, were without electricity. *Report of the Giant Power Survey Board to the General Assembly of the Commonwealth of Pennsylvania* (1925, 38).

39. Wells 1927, 14–15, 91–99.

40. Van Derzee 1928.

41. Parker 1913, 140.

42. Reed 1936, 759.

43. Rural Electrification Act, P. L. No. 74-605, 49 Stat. 1363 (1936); Roosevelt 1935.

44. Brown 1980, 35.

45. Ibid., 42.

46. "M'Carter Assails Curb on Utilities" (1935).

47. Quoted in Brown 1980, 40.

48. "Cheap Electricity Planned for Farm" (1935).

49. Brown 1980, 49.

50. "Utility Findings Are Called Fraud" (1935).

51. "Power Men Urged to Fight to Finish" (1935); "Utilities Get Call to 'Fight for Life'" (1935).

52. "Farm Power Plan Pushed at Capital" (1935).

53. A copy of the letter was submitted by Cooke to a congressional committee. *Rural Electrification* (1936, 34–38).

54. "Electric Service for Rural Areas" (1935); Person 1950, 74.

55. Richardson 1961, 25.

56. "Project Applications Received from 46 States" (1935).

57. Richardson 1961, 29.

58. Twentieth Century Fund 1948, 446.

59. Rural Electrification Act, P. L. No. 74-605, 49 Stat. 1363 (1936). Senator George W. Norris (R-NE) had been actively involved in the creation of the TVA (see chapter 7). Representative Sam Rayburn (D-TX) later became Speaker of the House.

60. Twentieth Century Fund 1948, 449.

61. Person 1950.

62. Tennessee Valley Authority 1934, 29.

63. Brown 1980, 36–38. Alcon's service area was not entirely rural. It included the towns of Corinth and Renzi, whose 1940 populations were about 6,000 and 500, respectively.

64. Morgan 1936, 796–97. Arthur Morgan was then a director of the TVA. In the text, he says there were thirty-four in operation, but the table he provided showed only thirty-three.

65. Tennessee Valley Authority 1936, 141, 289. The TVA did not provide contracts with distributors in their published annual reports prior to the one for the fiscal year 1936. The REA provided loans to eighteen TVA cooperatives in 1935. Some may have been in existence prior to the creation of the REA.

66. US Rural Electrification Administration 1938a, 10–11.

67. Person 1950, 70–79; Slattery and Mittell 1940, 57–63; Twentieth Century Fund 1948, 451–57; US Rural Electrification Administration 1938a, 51–57.

68. Brown 1980, 69–70; Slattery and Mittell 1940, 73.

69. Slattery and Mittell 1940, 38–48.

70. Ibid.; US Rural Electrification Administration 1938a, 43–51.

71. US Rural Electrification Administration 1938a, 58. See also *Agricultural Department Appropriation Bill for 1941* (1940, 1067–68).

72. These costs included "costs of primary and secondary distribution lines, substations, transformers, services, meters, inspection of poles and lines, engineering and legal services and all items of overhead attending the establishment of new enterprises such as costs of offices, warehouses, office equipment, trucks and an allowance for interest payments which accumulate until a project is energized." *Agricultural Department Appropriation Bill for 1941* (1940, 1069).

73. *Agricultural Department Appropriation Bill for 1941* (1940, 1067); Twentieth

Century Fund 1948, 453–54. These costs are not strictly comparable because the average number of customers per mile for the REA cooperatives was 2.2 while that for the privately owned utilities was 4.7. The amounts of some equipment used to distribute electricity depend on the number of customers, not the number of miles. However, the 1935 cost for the lines alone for the privately owned utilities was $910, while the total construction cost (substations, lines, transformers, and services) for REA cooperatives was $893, suggesting that the REA's costs were significantly lower when all factors were taken into account.

74. "Reductions in Insurance Rates Obtained for REA-Financed Cooperatives" (1938); *Agricultural Department Appropriation Bill for 1941* (1940, 1074–76); Slattery and Mittell 1940, 49–56; Twentieth Century Fund 1948, 454–57; US Rural Electrification Administration 1940, 83–84,103–24.

75. Twentieth Century Fund 1948, 468.

76. Richardson 1961, 87; US Rural Electrification Administration 1938a, 20–21.

77. "F. Harper Craddock Named Chief of REA Rate Section" (1938).

78. "Cut in Wholesale Rates in Indiana Opens Way for New REA Allotments" (1937); "Kentucky Commission Orders Low Rate for Wholesale Current" (1937); "Iowa Co-Ops Win Court Fight for Source of Wholesale Energy" (1938).

79. Twentieth Century Fund 1948, 447.

80. US Rural Electrification Administration 1940, 85, 88–89.

81. Twentieth Century Fund 1948, 459, 468–70.

82. Ibid., 464, 467. Although these data are the best available, they can only be considered approximate. The Twentieth Century Fund calculated the number of farms connected each year as the annual increment in total connected farms. The Edison Electric Institute published annual estimates of the total number of farms receiving service but did not separate the total by provider. Furthermore, those figures were periodically revised. Although the revisions were a small proportion of the total, they often had a larger effect on the year-to-year differences. This may be the source of the apparent year-to-year volatility, such as the sharp drop in 1938. Marple 1938. The REA provided data annually on the number of customers receiving service but did not provide the number of these customers that were farms. The Twentieth Century Fund assumed 80 percent of customers were farms, a percentage that had been used by the REA at other times.

83. The total number of customers of REA-financed projects that were receiving electricity in 1940 was 567,998. An additional 286,830 had not yet begun receiving electricity. If 80 percent of those receiving electricity were farms, the number of farms would be 454,398. During the decade ending in 1935, privately owned electric utilities had added an average of 54,245 farms per year. From 1936 to 1940, the number of farms receiving electricity from a privately owned utility increased by 649,766. This was 378,544 more farms than would have received service at the pre-1936 rate. Twentieth Century Fund 1948, 464, 467; US Rural Electrification Administration 1941, 28.

84. Porter 1938b; Twentieth Century Fund 1948, 470.

85. Nicholson 1937; Richardson 1961, 46–47; Slattery and Mittell 1940, 44.

86. Reed 1936, 757–58.

87. "REA Victory in Alabama Case Aids Fight against 'Spite Lines' in Other States" (1937, 30); US Rural Electrification Administration 1938a, 39. The REA submitted over 100 affidavits supporting these assertions during a hearing before a federal judge. Slattery and Mittell 1940, 116–17.

88. Richardson 1961, 63–79; Twentieth Century Fund 1948, 458–59.

89. US Rural Electrification Administration 1940, 152.

90. "REA Activities in Virginia Suspended" (1936); "REA Victory in Alabama Case Aids Fight against 'Spite Lines' in Other States" (1937); Cooke 1936; Richardson 1961, 63–79; Slattery and Mittell 1940, 117–20; US Rural Electrification Administration 1940, 153.

91. US Rural Electrification Administration 1938a, 36–37.

92. Slattery and Mittell 1940, 115–24.

93. "REA Victory in Alabama Case Aids Fight against 'Spite Lines' in Other States" (1937); US Rural Electrification Administration 1941, 11–12.

94. "P.S.C. Rulings Advance Wisconsin Rural Electrification" (1936); Twentieth Century Fund 1948, 458–59.

95. "Rate Clause Blocks Projects in the Northwest" (1936).

96. US Rural Electrification Administration 1938a, 100–105.

97. Bonbright 1940, 40.

98. "REA Victory in Alabama Case Aids Fight against 'Spite Lines' in Other States" (1937).

99. Kitchens and Fishback 2015.

100. Cowan 2014, 3.

101. US Rural Electrification Administration 1939, 101–4; US Rural Electrification Administration 1940, 142–43.

102. "Report of the Rural Lines Committee" (1922, 108–10).

103. In 1949, the interest rate charged by the REA was one-hundredth of a percentage point below the rate paid by the federal government. In the early 1980s, the REA's interest was over five percentage points below the federal government's rate. US Rural Electrification Administration 1985, 31.

104. Slattery and Mittell 1940, 81.

105. Dudley and Evans 1941.

106. Abrams 1940, 32–38; Reed 1936, 758–59.

107. Twentieth Century Fund 1948, 464–65.

108. Rural Electrification Administration 1964, 6.

109. This is the total allotments made by 1940 divided by the total number of customers. The latter came from estimates in the loan contracts and included those not yet receiving electricity. US Rural Electrification Administration 1941, 29-53.

110. *Agricultural Department Appropriation Bill for 1941* (1940, 1038).

111. Ibid., 1123.

112. The number of customers includes only those served by active borrowers

of the electric program who were receiving electricity. The 1940 number of customers included those in systems that had received loans but that were not yet energized. The REA adjusted the 1967 figures to account for loans cooperatives received from other sources. US Rural Electrification Administration 1968.

Conclusion and a Look Forward from 1940

1. This brief section seeks to outline some of the major events after 1940 that led to a radical restructuring of the industry in parts of the country.

2. Carter et al. 2006, table Db 228.

3. Sunk expenditures in one plant cancelled by Duke Power amounted to over 30 percent of the company's net worth. Hearth, Melicher, and Gurley 1990, 102.

4. Chen 1999; Sidak and Spulber 1997; Tomain 2002.

5. By 1991, the percentage of generation from outside the electric utility industry was 25 percent in California, 27 percent in Louisiana, and 20 percent in Texas. Gilbert and Kahn 1996, 204.

6. Hannah 1979, 100–49.

7. Green 1991.

8. An independent system operator (ISO) is very similar and performs the same functions. "RTO" will be used to refer to either type of organization.

9. Paying each supplier the same highest bid for electricity purchased creates an incentive for all suppliers to bid their marginal cost, which includes only operating costs, with the expectation of covering their capital costs with payments above their bid prices. This is closer to the way competitive markets work where different producers of identical products all will charge the same, or similar, prices even if their costs differ.

10. Borenstein and Bushnell 2000.

11. Borenstein 2002; Borenstein, Bushnell, and Wolak 2002; Joskow 2001.

12. Borenstein and Bushnell 2014, 6–8; Mansur and White 2012.

13. Green 2003; Hogan 2012; Joskow 2004; Overbye 2000.

14. Borenstein and Bushnell 2014, 13–18.

15. Much of the following material comes from Borenstein and Bushnell 2014.

16. The merger of Duke Power and Progress Energy created a single integrated utility serving most of both North and South Carolina, but the merged company also has other, nonadjacent, service areas.

17. The problem of market power has also been noted in other countries that have deregulated electricity generation. Borenstein and Bushnell 2000; Wolfram 1999.

18. Borenstein, Bushnell, and Wolak 2002.

19. Borenstein 2002, 198.

Bibliography

"$1,200 a Month Phone Graft for Boss Ruef." 1907. *The Atlanta Constitution*, March 31: B1.

"$45,500,000 of Water in New York Edison Co." 1905. *New York Times*, April 4: 2.

Abrams, Ernest Russell. 1940. *Power in Transition*. New York: C. Scribner's Sons.

Adams, Foster. 1936. "Rural Electrification." In *Transactions Third World Power Conference Vol. IX*, 475–93. Washington, DC: US Government Printing Office.

Agricultural Department Appropriation Bill for 1941. 1940. Hearings before US House Committee on Appropriations. December 4–11, 1939; January 9–13, 76th Congress, 3rd session.

Alchian, Armen. 1987. "Rent." In *The New Palgrave: A Dictionary of Economics*, edited by John Eatwell, Murray Milgate, Peter Newman, and Robert Harry Inglis Palgrave, 90-93. London: Macmillan.

"Alex Dow Denounces Speculation." 1925. *Electrical World* 86 (14): 709.

"Appeal from Rates." 1906. *Washington Post*, February 14: 5.

"Are the Federal Courts Unworthy of Confidence?" 1906. *New York Times*, February 22: 6.

Armstrong, Christopher, and H. V. Nelles. 1986. *Monopoly's Moment: The Organization and Regulation of Canadian Utilities, 1830–1930*. Philadelphia: Temple University Press.

Atkins, Leah Rawls. 2006. *"Developed for the Service of Alabama": The Centennial History of the Alabama Power Company, 1906–2006*. Birmingham: Alabama Power Co.

Averch, Harvey, and Leland L. Johnson. 1962. "Behavior of the Firm under Regulatory Constraint." *American Economic Review* 52 (5): 1052–69.

Barnard, Ellsworth. 1966. *Wendell Willkie: Fighter for Freedom*. Marquette: Northern Michigan University Press.

Barron, William J. 1942. "The Evolution of Smyth v. Ames: A Study of the Judicial Process." *Virginia Law Review* 28 (6): 761–94.

Baruch, Bernard M. 1941. *American Industry in the War: A Report of the War Industries Board (March 1921)*. New York: Prentice-Hall.

Bauer, John. 1926. "Reproduction Cost and Desirable Public Utility Regulation." *Journal of Land & Public Utility Economics* 2:408–26.

Bauer, John, and Nathaniel Gold. 1939. *The Electric Power Industry: Development, Organization, and Public Policies*. New York: Harper & Brothers.

Baumol, William J., and Alvin K. Klevorick. 1970. "Input Choices and Rate-of-Return Regulation: An Overview of the Discussion." *Bell Journal of Economics and Management Science* 1 (2): 162–90.

Belfield, Robert B. 1976. "The Niagara System: The Evolution of an Electric Power Complex at Niagara Falls, 1883–1896." *Proceedings of the IEEE* 64 (9): 1344–50.

Bellush, Bernard. 1955. *Franklin D. Roosevelt as Governor of New York*. New York: Columbia University Press.

Bemis, Edward W. 1927. "Going Value in Rate Cases in the Supreme Court." *Columbia Law Review* 27 (5): 530–46.

Berle, Adolf Augustus, and Gardiner C. Means. 1968. *The Modern Corporation and Private Property*. New York: Harcourt, Brace & World.

Billington, David P., Donald C. Jackson, and Martin V. Melosi. 2005. "The History of Large Federal Dams Planning, Design, and Construction in the Era of Big Dams." Denver: Bureau of Reclamation, US Department of the Interior. Accessed November 2, 2014. http://purl.access.gpo.gov/GPO/LPS102089.

Bills to Amend the Bonneville Act. 1942. Hearings before Senate Committee on Commerce and House Committee on Rivers and Harbors. June 3–19, 77th Congress, 2nd session.

Bonbright, James C. 1940. *Public Utilities and the National Power Policies*. New York: Columbia University Press.

Bonbright, James C., and Gardiner C. Means. 1932. *The Holding Company: Its Public Significance and Its Regulation*. New York: McGraw-Hill.

Bonneville Power Administration. 1939. "First Annual Report of the Bonneville Administrator," H. Doc 86, 76th Congress, 1st session.

———. 1940a. "Second Annual Report of the Administrator of the Bonneville Power Administration," H. Doc 612, 67th Congress, 3rd session.

———. 1940b. *Third Annual Report of the Administrator of the Bonneville Power Administration*. Portland, OR: US Department of the Interior.

Borenstein, Severin. 2002. "The Trouble with Electricity Markets: Understanding California's Restructuring Disaster." *Journal of Economic Perspectives* 16 (1): 191–211.

Borenstein, Severin, and James Bushnell. 2014. "The US Electricity Industry after 20 Years of Restructuring." Working paper, the Energy Institute at Haas, University of California, Berkeley. Accessed October 31, 2015. http://ei.haas.berkeley.edu/research/papers/WP252RR.pdf.

Borenstein, Severin, and James B. Bushnell. 2000. "Electricity Restructuring: Deregulation or Reregulation?" *Regulation* 23 (2): 46–52.

Borenstein, Severin, James B. Bushnell, and Frank A. Wolak. 2002. "Measuring Market Inefficiencies in California's Restructured Wholesale Electricity Market." *American Economic Review* 92 (5): 1376–405.

Bowers, Brian. 1982. *A History of Electric Light & Power.* Stevenage and New York: P. Peregrinus / The Science Museum.

———. 1998. *Lengthening the Day: A History of Lighting Technology.* Oxford and New York: Oxford University Press.

Brown, D. Clayton. 1970. "Rural Electrification in the South, 1920–1955." Ph.D. diss., University of California, Los Angeles.

———. 1980. *Electricity for Rural America: The Fight for the REA.* Westport, CT: Greenwood Press.

Bruere, Henry. 1908. "Public Utilities Regulation in New York." *Annals of the American Academy of Political and Social Science* 31:1–17.

Buchanan, Norman S. 1935. "The Capital Account and the Rate of Return in Public Utility Operating Companies." *Journal of Political Economy* 43:50–68.

———. 1936. "The Origin and Development of the Public Utility Holding Company." *Journal of Political Economy* 44:31–53.

Burness, H. S., R. G. Cummings, W. D. Gorman, and R. R. Lansford. 1980. "United States Reclamation Policy and Indian Water Rights." *Natural Resources Journal* 20 (4): 807–26.

Bushnell, S. M. 1909. "Central Station Operation of Steam Plants in Connection with Lighting Company's Service." In *National Electric Light Association, Thirty-Second Convention, Volume II.* 778–815. National Electric Light Association.

Cannon, Brian Q. 2000. "Power Relations: Western Rural Electric Cooperatives and the New Deal." *Western Historical Quarterly* 31 (2): 133–60.

Cannon, L. G. 1934. "The Future of the Electric Utility Industry: Fated or Planned?" *Journal of Land & Public Utility Economics* 10 (2): 150–66.

Carlson, W. Bernard. 1991. *Innovation as a Social Process: Elihu Thomson and the Rise of General Electric, 1870–1900.* Cambridge and New York: Cambridge University Press.

Caro, Robert A. 1982. *The Years of Lyndon Johnson: The Path to Power.* New York: Vintage Books.

Carter, F. B. 1909. "Isolated Plants." In *National Electric Light Association, Thirty-Second Annual Convention, Volume II,* 754–64. National Electric Light Association.

Carter, Susan B., Scott Sigmund Gartner, Michael R. Haines, Alan L. Olmstead, Richard Sutch, and Gavin Wright. 2006. "Historical Statistics of the United States Millennial Edition Online," Cambridge University Press. http://hsus.cambridge .org/.

"Chattanooga Vote Is for T.V.A. Power." 1935. *New York Times,* March 13: 4.

Chazeau, Melvin G. de. 1934a. "I. Rationalization of Electricity Supply in Great Britain." *Journal of Land & Public Utility Economics* 10 (3): 254–67.

———. 1934b. "II. The Rationalization of Electricity Supply in Great Britain." *Journal of Land & Public Utility Economics* 10 (4): 365–90.

"Cheap Electricity Planned for Farm." 1935. *New York Times*, May 14: 27.

Chen, Jim. 1999. "The Second Coming of Smyth v. Ames." *Texas Law Review* 77 (6): 1535.

Christie, Jean. 1972. "Giant Power: A Progressive Proposal of the Nineteen-Twenties." *Pennsylvania Magazine of History and Biography* 96 (4): 480–507.

Cintrón, Lizette, and Edison Electric Institute. 1995. *Historical Statistics of the Electric Utility Industry, 1902–1992*. Washington, DC: Edison Electric Institute.

"Colorado River Compact." 1922. Accessed July 20, 2015. http://www.usbr.gov/lc/region/pao/pdfiles/crcompct.pdf.

"Columbia River and Minor Tributaries, Vol I." 1933. H. R. Doc 103, 73rd Congress, 1st session.

"Columbia River and Minor Tributaries, Vol II." 1933. H. R. Doc 103, 73rd Congress, 1st session.

"Comments on Barstow." 1898. In *1898 AEIC Minutes*, 33–35, 125–35. New York: Association of Edison Illuminating Companies.

Commons, John R. 1905. "The Lafollette Railroad Law in Wisconsin." *American Monthly Review of Reviews*, July: 76–79.

———. 1934. *Myself*. New York: Macmillan.

Commonwealth and Southern Corporation. 1937. *Analysis of the Annual Report of the Tennessee Valley Authority, Released on December 31, 1936*. New York: Commonwealth and Southern Corporation.

Commonwealth Edison. 1909. "Electric Light from Station is Cheapest." Advertisement. Reproduced and discussed in *Isolated Plant* 1, no. 5 (August): 12. Accessed May 22, 2013. http://google.com/books?id=WwwKAAAAIAAJ.

Conservation and Development of the National Resources. 1937. Hearings before House Committee on Rivers and Harbors. 75th Congress, 1st session.

Consolidated Edison. 2007. "A/C but No D/C: Last Con Edison Direct Current Customer Is History." Accessed Nov. 12, 2012. http://www.coned.com/newsroom/news/pr20071115.asp.

"Control of Franchises." 1903. *New York Times*, February 27: 16.

Cooke, Morris L. 1925. "A Note on Rates for Rural Electric Service." *Annals of the American Academy of Political and Social Science* 118:52–59.

———. 1936. "Plan or Muddle Through." *Rural Electrification News* 1 (9): 3–4.

Coppock, Joseph D. 1940. *Organization and Operations of the Electric Home and Farm Authority*. Cambridge, MA: National Bureau of Economic Research. Accessed June 5, 2014. http://www.nber.org/chapters/c4943.pdf.

Cowan, Tadlocki. 2014. "An Overview of USDA Rural Development Programs." Congressional Research Service RL31837.

Cowles, Alfred, III, and Associates. 1939. *Common-Stock Indexes, 1871–1937*. 2nd ed. Bloomington, IN: Principia Press.

Creamer, Daniel Barnett, Sergei P. Dobrovolsky, and Israel Borenstein. 1960. *Capital in Manufacturing and Mining: Its Formation and Financing*. Princeton: Princeton University Press.

Crickmer, Barry. 1993. "Edison Electric Institute the First 60 Years." *Electric Perspectives* 17 (3): 46.

Cushman, Robert Eugene. 1972. *The Independent Regulatory Commissions*. New York: Octagon Books.

"Cut in Wholesale Rates in Indiana Opens Way for New REA Allotments." 1937. *Rural Electrification News* 3 (1): 18.

David, Paul A. 1991. "The Hero and the Herd in Technological History: Reflections on Thomas Edison and the Battle of the Systems." In *Favorites of Fortune: Technology, Growth, and Economic Development since the Industrial Revolution*, edited by Patrice Higonnet, David Landes, and Henry Rosovsky, 72–119. Cambridge, MA: Harvard University Press.

David, Paul A., and Julie Ann Bunn. 1988. "The Economics of Gateway Technologies and Network Evolution: Lessons from Electricity Supply History." *Information Economics and Policy* 3:165–202.

Davis, Ernest H. 1905. "Presidential Address." In *National Electric Light Association, Twenty-Eighth Convention, Volume I*, 6–11. National Electric Light Association.

Davis, G. Cullom. 1962. "The Transformation of the Federal Trade Commission, 1914–1929." *Mississippi Valley Historical Review* 49 (3): 437–55.

Davis, Kenneth S. 1985. *FDR: The New York Years, 1928–1933*. New York: Random House.

DeGraaf, Leonard. 1990. "Corporate Liberalism and Electric Power System Planning in the 1920s." *Business History Review* 64:1–31.

Delco. 1923. "Ad for Isolated Plants for Farms." *Farm Mechanics*, 15.

Denayrouze, L., and Jablochkoff. 1877. "Divisibilité De La Lumière Électrique." *Comptes Rendus Hebdomadaires des Séances de l'Académie des Sciences* 84: 750–52.

Devine, Warren D. 1983. "From Shafts to Wires: Historical Perspective on Electrification." *Journal of Economic History* 43 (2): 347–72.

Dillavou, E. R. 1933. "Desirable Legal Changes in Holding Company Legislation." *Accounting Review* 8 (1): 43–50.

"Discussion of Papers on the Issue of Central Stations vs Isolated Plants." 1910. *Proceedings of the American Institute of Electrical Engineers and the American Society of Mechanical Engineers* 29:977–1009.

"Discussion on Interconnection and Superpower." 1924. In *1924 AEIC Minutes*, 213–30. New York: Association of Edison Illuminating Companies.

Doherty, Henry L. 1902. "Presidential Address." In *National Electric Light Association, Twenty-Fifth Convention*, 28-45. New York: James Kempster Printing Company.

Dorau, Herbert B. 1929. *The Changing Character and Extent of Municipal Ownership in the Electric Light and Power Industry*. Chicago: The Institute for Research in Land Economics and Public Utilities.

———. 1930. *Materials for the Study of Public Utility Economics.* New York: Macmillan.

Du Boff, Richard B. 1964. "Electric Power in American Manufacturing, 1889–1958." Ph.D. diss., University of Pennsylvania.

Dudley, R. G., and W. H. Evans. 1941. "Public Utility Financing in the Fourth Quarter of 1940 and a Summary of the Year, 1940." *Journal of Land & Public Utility Economics* 17 (1): 118–24.

Edison, Thomas A. 1883. System of Electrical Distribution. US Patent 274,290, filed November 27, 1882, and issued March 20, 1883.

"Electrical Plant in the Newark Free Public Library." 1903. *Electrical World and Engineer* 42:271–72.

"Electric Light Costs Small Consumer Dear." 1905. *New York Times*, April 7: 3.

"Electric Rates-Massachusetts." 1912. *Rate Research* 2 (4): 48–53.

"Electric Service for Rural Areas." 1935. *New York Times*, August 1: 21.

Ely, Owen. 1936. "The 'Power Pool' Proposal." *Public Utilities Fortnightly* 18: 679–87.

Emergency Power Bill Hearings. 1918. Hearings before US Senate Committee on Commerce. October 11–19, 65th Congress, 2nd session.

Evans, Gale. 1992. "Storm over Niagara: A Catalyst in Reshaping Government in the United States and Canada During the Progressive Era." *Natural Resources Journal* 32 (1): 27–54.

Fall, Albert, and A. P. Davis. 1922. "Development of the Imperial Valley," S. Doc. 142, 67th Congress, 2nd session.

"Farm Power Plan Pushed at Capital." 1935. *New York Times*, July 11: 3.

Federal Columbia River Power System. 2001. *The Columbia River System: Inside Story.* Portland, OR: Bonneville Power Administration.

"F. Harper Craddock Named Chief of REA Rate Section." 1938. *Rural Electrification News* 3 (5): 25.

Field, Gregory B. 1990. "'Electricity for All': The Electric Home and Farm Authority and the Politics of Mass Consumption, 1932–1935." *Business History Review* 64:32–60.

Field, Kenneth. 1932. "Holding Corporation Control as a Provisional Form of Consolidation." *Journal of Land & Public Utility Economics* 8 (1): 87–96.

Fishback, Price V., William C. Horrace, and Shawn Kantor. 2005. "Did New Deal Grant Programs Stimulate Local Economies? A Study of Federal Grants and Retail Sales During the Great Depression." *Journal of Economic History* 65 (1): 36–71.

Fisher, Walter L. 1907. "The American Municipality." In *Municipal and Private Operation of Public Utilities: Report to the National Civic Federation*, 1:33–42. New York: National Civic Federation.

"Fisher Says Prices of Stocks Are Low." 1929. *New York Times*, October 22: 24.

Fleming, Keith Robson. 1992. *Power at Cost: Ontario Hydro and Rural Electrification, 1911–1958.* Montreal: McGill-Queen's University Press.

Freeman, Neil B. 1996. *The Politics of Power: Ontario Hydro and Its Government, 1906–1995*. Toronto: University of Toronto Press.

Friedel, Robert D., Paul Israel, and Bernard S. Finn. 1986. *Edison's Electric Light: Biography of an Invention*. New Brunswick, NJ: Rutgers University Press.

Funigiello, Philip J. 1973. *Toward a National Power Policy: The New Deal and the Electric Utility Industry, 1933–1941*. Pittsburgh: University of Pittsburgh Press.

Garcke, Emile. 1911. "Electricity Supply—II Commercial Aspects." *Encyclopædia Britannica*, 13th ed.

"Gas Report Held up by Tammany Panic." 1905. *New York Times*, April 30: 3.

Gear, Harry Barnes, and Paul Francis Williams. 1911. *Electric Central Station Distribution Systems, Their Design and Construction*. New York: D. Van Nostrand Company.

"Geographic Integration under the Public Utility Holding Company Act." 1941. *Yale Law Journal* 50 (6): 1045–55.

Gilbert, Richard J., and Edward P. Kahn. 1996. "Competition and Institutional Change in US Electric Power Regulation." In *International Comparisons of Electricity Regulation*, edited by Richard J. Gilbert and Edward P. Kahn.: 179–230. New York: Cambridge University Press.

Glaeser, Martin G. 1929. *Outlines of Public Utility Economics*. New York: Macmillan.

Goldberg, Victor P. 1976. "Regulation and Administered Contracts." *Bell Journal of Economics* 7 (2): 426–48.

Gray, John H. 1900. "The Gas Commission of Massachusetts." *Quarterly Journal of Economics* 14 (4): 509–36.

Gray, Rowena. 2013. "Taking Technology to Task: The Skill Content of Technological Change in Early Twentieth Century United States." *Explorations in Economic History* 50 (3): 351–67.

"Great Southern Transmission Network, The." 1914. *Electrical World*, May 30: 1201.

Green, Richard. 1991. "Reshaping the CEGB: Electricity Privatization in the UK." *Utilities Policy* 1 (3): 245–54.

———. 2003. "Cost Recovery and the Efficient Development of the Grid." In *Transport Pricing of Electricity Networks*, edited by François Lévêque, 137–53. Boston: Kluwer Academic Publishers.

Greenwood, Ernest. 1928. *Aladdin, USA*. New York: Harper & Brothers.

Gressley, Gene M. 1968. "Arthur Powell Davis, Reclamation, and the West." *Agricultural History* 42 (3): 241–57.

Grossman, Richard S. 1992. "Deposit Insurance, Regulation, and Moral Hazard in the Thrift Industry: Evidence from the 1930s." *American Economic Review* 82 (4): 800–821.

Gruening, Ernest. 1964 [1931]. *The Public Pays: A Study of Power Propaganda*. New York: Vanguard Press.

Gruhl, Edwin. 1914. "Recent Tendencies in Valuations for Rate-Making Purposes." *Annals of the American Academy of Political and Social Science* 53:219–37.

Hagenah, William J. 1932. "Influence of Reproduction Cost in Electric Utility

Valuations." In *1932 AEIC Minutes*, 323–43. New York: Association of Edison Illuminating Companies.

Hale, R. S. 1903. "Isolated Plant vs. Central Stations Supply of Electricity: A Suggestion for Obtaining Estimates of Costs on a Competitive Basis." *Electrical World and Engineer* 42:383–84.

———. 1910. "The Supply of Electrical Power for Industrial Establishments from Central Stations." *Proceedings of the American Institute of Electrical Engineers and the American Society of Mechanical Engineers* 29:219–27.

Hannah, Leslie. 1979. *Electricity before Nationalization: A Study of the Development of the Electricity Supply Industry in Britain to 1948*. Baltimore: Johns Hopkins University Press.

Harbeson, Robert W. 1936. "The Power Program of the Tennessee Valley Authority." *Journal of Land & Public Utility Economics* 12 (1): 19–32.

Hausman, William J., and John L. Neufeld. 1984. "Time-of-Day Pricing in the US Electric Power Industry at the Turn of the Century." *Rand Journal of Economics* 15:116–26.

———. 1990. "The Structure and Profitability of the US Electric Utility Industry at the Turn of the Century." *Business History* 32:225–43.

———. 1991. "Property Rights Versus Public Spirit: Ownership and Efficiency of US Electric Utilities Prior to Rate-of-Return Regulation." *Review of Economics and Statistics* 73 (3): 414–23.

———. 1992. "Battle of the Systems Revisited: The Role of Copper." *IEEE Technology and Society Magazine* 11 (3): 18–25.

———. 1994. "Public Versus Private Electric Utilities in the United States: A Century of Debate over Economic Efficiency." *Annals of Public and Cooperative Economics* 65 (4): 599–622.

———. 2002. "The Market for Capital and the Origins of State Regulation of Electric Utilities in the United States." *Journal of Economic History* 62 (4): 1050–73.

Healy, Robert E. 1936. "Organization of Private Electric and Gas Utilities." Paper presented at the Third World Power Conference. Vol. V, 279–329.

Hearth, Douglas, Ronald W. Melicher, and Darryl E. J. Gurley. 1990. "Nuclear Power Plant Cancellations: Sunk Costs and Utility Stock Returns." *Quarterly Journal of Business and Economics* 29 (1): 102–15.

Heilman, Ralph E. 1914. "The Development by Commissions of the Principles of Public Utility Valuation." *Quarterly Journal of Economics* 28 (2): 269–91.

———. 1925. "Customer Ownership of Public Utilities." *Journal of Land & Public Utility Economics* 1 (1): 7–17.

Hogan, William W. 2012. "Multiple Market-Clearing Prices, Electricity Market Design and Price Manipulation." Working paper, John F. Kennedy School of Government, Harvard University. Accessed October 31, 2014. http://www.hks.harvard.edu/fs/whogan/Hogan_Degenrate_Price_033112r.pdf.

Homan, Paul T. 1931. "Economic Aspects of the Boulder Dam Project." *Quarterly Journal of Economics* 45 (2): 177–217.

Hon, Ralph C. 1940. "The Memphis Power and Light Deal." *Southern Economic Journal* 6 (3): 344–75.

Hoover, Herbert. 1931. "Veto Message Relating to Disposition of Muscle Shoals." S. Doc. 321, 71st Congress, 3rd session.

Hormell, Orren C. 1932. "Ownership and Regulation of Electric Utilities in Great Britain." *Annals of the American Academy of Political and Social Science* 159: 128–39.

Hubbard, Preston J. 1961. *Origins of the TVA: The Muscle Shoals Controversy, 1920–1932*. Nashville, TN: Vanderbilt University Press.

Hughes, Jonathan R. T. 1986. *The Vital Few: The Entrepreneur and American Economic Progress*. New York: Oxford University Press.

Hughes, Thomas P. 1958. "Harold P. Brown and the Executioner's Current: An Incident in the AC-DC Controversy." *Business History Review* 32:143–65.

———. 1983. *Networks of Power: Electrification in Western Society, 1880–1930*. Baltimore: Johns Hopkins University Press.

Hunter, Louis C., and Linwood Bryant. 1991. *A History of Industrial Power in the United States, 1780–1930, Volume 3: The Transmission of Power*. Cambridge, MA: MIT Press.

Insull, Samuel. 1898. "Presidential Address." In *National Electric Light Association, Twenty-First Convention*, 14–30. Accessed December 15, 2015. https://books.google.com/books?id=iW9NAAAAYAAJ.

———. 1924. *Public Utilities in Modern Life: Selected Speeches (1914–1923)*. Chicago: privately printed.

Insull, Samuel, and Larry Plachno. 1992. *The Memoirs of Samuel Insull*. Polo, IL: Transportation Trails.

"Insull Rose to Top of Utilities Empire." 1938. *New York Times*, July 17: 26.

Investigation of Public Utility Corporations, Part 2. 1928. Hearings before US Senate Committee on Interstate Commerce. January 17–18. 70th Congress, 1st session.

Investigation of the Tennessee Valley Authority, Part 10. 1938. Hearings before US Joint Committee on the Investigation of the Tennessee Valley Authority. November 22–29, 75th Congress, 3rd session.

"Iowa Co-Ops Win Court Fight for Source of Wholesale Energy." 1938. *Rural Electrification News*, 23.

"Isolated Plant Versus the Central Station, The—Discussion at the Chicago Electrical Association." 1902. *Electrical World and Engineer* 39 (19): 817–19.

"Isolated Plants—Steam Heating." 1910. *Electrical World* 55:1279.

James, George Wharton. 1917. *Reclaiming the Arid West: The Story of the United States Reclamation Service*. New York: Dodd Mead and Company.

Jensen, Gordon Maurice. 1956. "The National Civic Federation: American Business in an Age of Social Change and Social Reform". Ph.D. diss., Princeton University.

Joint Committee on the Investigation of the Tennessee Valley Authority. 1939. "Report of the Joint Committee, Appendix A," S. Doc. 56, Part 2, 76th Congress, 1st session.

Jones, Eliot, and Truman C. Bigham. 1931. *Principles of Public Utilities.* New York: Macmillan.

Jones, William K. 1979. "Origins of the Certificate of Public Convenience and Necessity: Developments in the States, 1870–1920." *Columbia Law Review* 79 (3): 426–516.

Joskow, Paul L. 2001. "California's Electricity Crisis." *Oxford Review of Economic Policy* 17 (3): 365–88.

———. 2004. "Transmission Policy in the United States." Cambridge Working Papers in Economics, CMI Working Paper 54, University of Cambridge.

"J. Pierpont Morgan." 2014. NNDB. Accessed May 15, 2014. http://www.nndb.com /people/965/000082719/.

Junkersfeld, P. 1903. "Some Problems in the Design, Development and Operation of Large Central Station Systems." In *1903 AEIC Minutes*, 254–65. New York: Association of Edison Illuminating Companies.

Kahl, William L. 1983. *Water and Power: The Conflict over Los Angeles Water Supply in the Owens Valley.* Los Angeles: University of California Press.

Kapp, Gisbert. 1892. "Letter to the Editor on Price of Electric Supply." *Electrician* 30:199–200.

Keene, E. S. 1918. *Mechanics of the Household: A Course of Study Devoted to Domestic Machinery and Household Mechanical Appliances.* New York: McGraw-Hill Book Co.

Kendrick, John W. 1961. *Productivity Trends in the United States.* Princeton: Princeton University Press.

"Kentucky Commission Orders Low Rate for Wholesale Current." 1937. *Rural Electrification News* 2 (10): 24–25.

Kerwin, Jerome G. 1926. *Federal Water-Power Legislation.* New York: Columbia University Press.

King, W. James. 1962. "The Development of Electrical Technology in the 19th Century: 3. The Early Arc Light and Generator." *Contributions from the Museum of History and Technology, Smithsonian Institution* Bulletin 228: 333–407.

Kitchens, Carl. 2013a. "A Dam Problem: TVA's Fight against Malaria, 1926–1951." *Journal of Economic History* 73 (3): 694–724.

———. 2013b. "The Use of Eminent Domain in Land Assembly: The Case of the Tennessee Valley Authority." *Public Choice*: 1–12. Accessed August 14. doi:10.1007/s11127–013–0102-x.

———. 2014. "The Role of Publicly Provided Electricity in Economic Development: The Experience of the Tennessee Valley Authority, 1929–1955." *Journal of Economic History* 74:389–419.

Kitchens, Carl, and Price Fishback. 2015. "Flip the Switch: The Spatial Impact of the Rural Electrification Administration, 1935–1940." *Journal of Economic History* 75:1161–95.

Klein, Benjamin, Robert G. Crawford, and Armen A. Alchian. 1978. "Vertical Integration, Appropriable Rents, and the Competitive Contracting Process." *Journal of Law and Economics* 21 (2): 297–326.

Kleinsorge, Paul Lincoln. 1941. *The Boulder Canyon Project, Historical and Economic Aspects*. Redwood City, CA: Stanford University Press.

Kline, Patrick, and Enrico Moretti. 2014. "Local Economic Development, Agglomeration Economies, and the Big Push: 100 Years of Evidence from the Tennessee Valley Authority." *Quarterly Journal of Economics* 129 (1): 275–331.

Knittel, Christopher R. 2006. "The Adoption of State Electricity Regulation: The Role of Interest Groups." *Journal of Industrial Economics* 54 (2): 201–22.

Knowlton, H. S. 1907. "The Central Station and the Isolated Plant." *Cassier's* 32: 359–63.

"Knoxville Votes Municipal Power." 1933. *New York Times*, November 26: 27.

Krock, Arthur. 1936. "TVA Power Pool Plan Threatened by Attack." *New York Times*, December 20: E3.

Kruesi, Frank E., and Edwin Vennard. 1936. "Severance Damages Relating to Electric Utilities." *Journal of Land & Public Utility Economics* 12 (4): 361–74.

Layson, Stephen K. 1994. "Market Opening under Third-Degree Price Discrimination." *Journal of Industrial Economics* 42 (3): 335–40.

Learned, John G., E. P. Edwards, J. E. Schuff, S. M. Kennedy, J. Lukes, H. L. Montgomery, Herman Russell, et al. 1911. "Report of Committee on Electricity in Rural Districts." In *National Electric Light Association, Thirty-Fourth Convention, Volume I*, 448–534. National Electric Light Association.

Leatherwood, E. O. 1928. "My Objections to the Boulder Dam Project." *Annals of the American Academy of Political and Social Science* 135:133–40.

Leibenstein, Harvey. 1966. "Allocative Efficiency vs. 'X-Efficiency.'" *American Economic Review* 56 (3): 392–415.

Leuchtenburg, William E. 1952. "Roosevelt, Norris and the 'Seven Little TVAs.'" *Journal of Politics* 14 (3): 418–41.

Levin, Jack. 1931. *Power Ethics: An Analysis of the Activities of the Public Utilities in the United States, Based on a Study of the US Federal Trade Commission Records*. New York: A. A. Knopf.

Levy, Brian, and Pablo T. Spiller. 1994. "The Institutional Foundations of Regulatory Commitment: A Comparative Analysis of Telecommunications Regulation." *Journal of Law, Economics, & Organization* 10 (2): 201–46.

Libecap, Gary D. 2005. *Rescuing Water Markets: Lessons from Owens Valley*. Bozeman, Montana: PERC.

———. 2007. "The Assignment of Property Rights on the Western Frontier: Lessons for Contemporary Environmental and Resource Policy." *Journal of Economic History* 67 (2): 257–91.

Lilienthal, David E. 1964. *The TVA Years, 1939–1945*. New York: Harper & Row.

"Lobby Inquiry, The: Mr. Hopson Loses at Hide and Seek." 1935. *New York Times*, August 18: 1.

Lowitt, Richard. 1968. "Ontario Hydro: A 1925 Tempest in an American Teapot." *Canadian Historical Review* 49 (3): 267–74.

———. 1971. *George W. Norris, the Persistence of a Progressive: 1913–1933*. Urbana: University of Illinois Press.

Lyon, Thomas P., and Nathan Wilson. 2012. "Capture or Contract? The Early Years of Electric Utility Regulation." *Journal of Regulatory Economics* 42 (3): 225–41.

Magnusson, C. Edward. 1937. "Three-in-One Electric-Power Utilities and Regional Power Grids." In *Completion, Maintenance and Operation of Bonneville Project, Appendix A, Hearings before House Committee on Rivers and Harbors, March, April, May, and June*, 482–88. 75th Congress, 1st session.

Main, Charles T. 1910. "Central Stations Versus Isolated Plants for Textile Mills." *Proceedings of the American Institute of Electrical Engineers and the American Society of Mechanical Engineers* 29:205–17.

Maltbie, Milo R. 1912. "Judicial Review of Public Regulation." *Journal of Political Economy* 20 (5): 480–91.

Mansur, Erin T., and Matthew W. White. 2012. "Market Organization and Efficiency in Electricity Markets." Hanover, NH: Dartmouth College. Accessed June 3, 2015. http://www.dartmouth.edu/~mansur/papers/mansur_white_pjmaep.pdf.

Marlett, D. L., and Warren H. Marple. 1935. "Public Utility Legislation in the Depression: III. State Laws of 1935." *Journal of Land & Public Utility Economics* 11 (4): 390–99.

Marlett, D. L., and Orba F. Traylor. 1935a. "Public Utility Legislation in the Depression: II. Reorganization and Financing of Commissions." *Journal of Land & Public Utility Economics* 11 (3): 290–301.

———. 1935b. "Public Utility Legislation in the Depression: I. State Laws Extending and Strengthening Commission Jurisdiction." *Journal of Land & Public Utility Economics* 11 (2): 173–86.

Marple, Warren H. 1938. "An Appraisal of Edison Electric Institute's Statistics on Farm Electrification." *Journal of Land & Public Utility Economics* 14 (4): 471–76.

Martin, T. C. 1914. "Report of the Committee on Progress." Paper presented at the Thirty-Seventh Convention of the National Electric Light Association, Philadelphia, June 1–5, 146–49.

———. 1915. "Report of the Committee on Progress." Paper presented at the Thirty-Eighth Convention of the National Electric Light Association, Chicago, 152–213.

———. 1916. "Report of the Committee on Progress." Paper presented at the Thirty-Ninth Convention of the National Electric Light Association, San Francisco, June 7–11, 117–27.

Maxwell, J. T. 1909. "Isolated Plants." In *Thirty-Second Convention of the National Electric Light Association*, 764–77. Atlantic City: National Electric Light Association.

"M'Carter Assails Curb on Utilities." 1935. *New York Times*, May 9: 31.

McCollester, Parker. 1922. "Regulation of Intrastate Commerce under the Commerce Clause." *Yale Law Journal* 31 (8): 870–79.

McCormick, Richard L. 1981. "The Discovery That Business Corrupts Politics: A Reappraisal of the Origins of Progressivism." *American Historical Review* 86 (2): 247–74.

McCraw, Thomas K. 1970. *Morgan vs. Lilienthal: The Feud within the TVA*. Chicago: Loyola University Press.

———. 1971. *TVA and the Power Fight, 1933–1939*. Philadelphia: J. B. Lippincott Co.

McCrory, S. H. 1930. "Problems Involved and Methods Used in Promoting Rural Electrification." *Journal of Farm Economics* 12 (2): 320–25.

McDonald, Forrest. 1958. "Samuel Insull and the Movement for State Utility Regulatory Commissions." *Business History Review* 32:241–54.

———. 1962. *Insull*. Chicago: University of Chicago Press.

McGeary, M. Nelson. 1960. *Gifford Pinchot: Forester Politician*. Princeton: Princeton University Press.

McGrattan, Ellen R., and Edward C. Prescott. 2004. "The 1929 Stock Market: Irving Fisher Was Right." *International Economic Review* 45 (4): 991–1009.

McGuire, Patrick, and Marc Granovetter. 1998. "Business and Bias in Public Policy Formation: The National Civic Federation and Social Construction of Electric Utility Regulation, 1905–1907." Paper presented at the Annual Meeting of the American Sociological Association, San Francisco, CA, August.

McLean, Simon J. 1900. "State Regulation of Railways in the United States." *Economic Journal* 10 (39): 349–69.

Middle West Utilities Company. 1930. *Harvests and Highlines*. Chicago: Middle West Utilities Company.

Miller, George H. 1971. *Railroads and the Granger Laws*. Madison: University of Wisconsin Press.

Mills, J. Warner. 1905. "The Economic Struggle in Colorado." *Arena* 34 (192): 485–95.

Mitchell, Sidney Alexander. 1960. *S. Z. Mitchell and the Electrical Industry*. New York: Farrar, Straus & Cudahy.

Moeller, Beverley Bowen. 1971. *Phil Swing and Boulder Dam*. Berkeley: University of California Press.

Moffett, Edward A. 1907. "Introduction." In *Municipal and Private Operation of Public Utilities*, vol. 1, 12–19. New York: National Civic Federation.

Morgan, Harcourt A. 1936. "Rural Electrification: A Promise to American Life." In *Transactions of the Third World Power Conference*, 769–803. Washington, DC: Third World Power Conference.

———. 1938. "Investment of the Tennessee Valley Authority in Wilson, Norris and Wheeler Projects," H. R. Doc. 709, 75th Congress, 3rd session.

Morrison, Stephen A., and Clifford Winston. 2000. "The Remaining Role for Government Policy in the Deregulated Airline Industry." In *Deregulation of Network Industries: What's Next*, edited by Sam Peltzman and Clifford Winston., 1–40 Washington, DC: AEI-Brookings Joint Center for Regulatory Policy.

Mosher, William E. 1962. "Public Utilities and Their Recent Regulation." In *History of the State of New York*, edited by Alexander C. Flick. Vol. 8, *Wealth and Commonwealth*, 233–70. Port Washington, NY: Ira J. Friedman, Inc.

Mosher, William E., et al. 1929. *Electrical Utilities: The Crisis in Public Control.* New York: Harper & Brothers.

Mosher, William E., and Finla G. Crawford. 1933. *Public Utility Regulation.* New York: Harper & Brothers.

Mowry, George Edwin. 1951. *The California Progressives.* Berkeley: University of California Press.

"Municipal Ownership Convention, The." 1903. *Arena* 29 (5): 473.

"Municipal Ownership Convention Meets." 1903. *New York Times*, February 26: 6.

Murray, William Spencer. 1925. *Superpower: Its Genesis and Future.* New York: McGraw-Hill.

Murray, W. S., et al. 1921. *A Superpower System for the Region between Boston and Washington.* Washington, DC: US Government Printing Office.

Nash, L. R. 1933. *Public Utility Rate Structures: A Reference Book for Rate Designers, Executives, and Students.* New York: McGraw-Hill.

"Nation to Be Dark One Minute Tonight after Edison Burial." 1931. *New York Times* October 21: 1.

National Civic Federation. 1907a. *Municipal and Private Operation of Public Utilities: Report to the National Civic Federation Commission on Public Ownership and Operation.* Part I–Volume I: *General Conclusions and Reports.* New York: National Civic Federation.

———. 1907b. *Municipal and Private Operation of Public Utilities: Report to the National Civic Federation Commission on Public Ownership and Operation.* Part II-Volume I: *Report of Experts—United States.* New York: National Civic Federation.

———. 1907c. *Municipal and Private Operation of Public Utilities: Report to the National Civic Federation Commission on Public Ownership and Operation.* Part II-Volume II: *Report of Experts—United Kingdom.* New York: National Civic Federation.

———. 1913. *Commission Regulation of Public Utilities.* New York: National Civic Federation.

National Electric Light Association. 1932. *Progress in Rural and Farm Electrification for the 10 Year Period 1921–1931.* New York: National Electric Light Association.

Neil, Charles E. 1942. "Entering the Seventh Decade of Electric Power." *Edison Electric Bulletin* 10:321–32.

Neufeld, John L. 1987. "Price Discrimination and the Adoption of the Electricity Demand Charge." *Journal of Economic History* 47:693–709.

———. 2008. "Corruption, Quasi-Rents, and the Regulation of Electric Utilities." *Journal of Economic History* 68 (4): 1059–97.

Neufeld, John L., William J. Hausman, and Ronald B. Rapoport. 1994. "A Paradox of Voting: Cyclical Majorities and the Case of Muscle Shoals." *Political Research Quarterly* 47 (2): 423–38.

"New Civic Federation, A." 1899. *New York Times*, September 25: 1.

Newton, F. A. 1935. "The Commonwealth and Southern Objective Rate Plan." *Journal of Land & Public Utility Economics* 11 (2): 117–22.

Nicholson, Vincent. 1937. "Legal Steps Outlined for REA Projects." *Rural Electrification News* 2 (12): 21–22.

Nitrate Division, Ordinance Office War Department. 1922. *Report on the Fixation and Utilization of Nitrogen*. Washington, DC: US Government Printing Office.

Ogden, Daniel Miller, Jr. 1949. "The Development of Federal Power Policy in the Pacific Northwest." Ph.D. diss., University of Chicago.

Overbye, Thomas J. 2000. "Reengineering the Electric Grid." *American Scientist* 88:220–29.

Owen, Marguerite. 1973. *The Tennessee Valley Authority*. New York: Praeger.

Oxley, E. 1897. Multiple Metering of Electric Currents. US Patent 593,852, filed October 2, 1897, and issued November 16, 1897.

Parker, John C. 1913. "Report of the Committee on Electricity on the Farm." In *National Electric Light Association, Thirty-Sixth Convention, Commercial Sessions*, 137–297. National Electric Light Association.

Parsons, Frank. 1907. "The National Civic Federation and Its New Report on Public-Ownership." *Arena* 38 (215): 401–8.

Passer, Harold C. 1953. *The Electrical Manufacturers, 1875–1900*. Cambridge: Harvard University Press.

Pecora, Ferdinand. 1939. *Wall Street under Oath: The Story of Our Modern Money Changers*. New York: Simon and Schuster.

Peltzman, Sam. 1971. "Pricing in Public and Private Enterprises: Electric Utilities in the United States." *Journal of Law & Economics* 14 (1): 109–47.

———. 1976. "Toward a More General Theory of Regulation." *Journal of Law and Economics* 19 (2), Conference on the Economics of Politics and Regulation: 211–40.

Person, H. S. 1950. "The Rural Electrification Administration in Perspective." *Agricultural History* 24 (2): 70–89.

Pinchot, Gifford. 1925. "Governor Pinchot's Message of Transmittal." In *Report of the Giant Power Survey Board to the General Assembly of Commonwealth of Pennsylvania*, iii–xiii. Harrisburg, PA.

Pisani, Donald J. 1992. *To Reclaim a Divided West: Water, Law, and Public Policy, 1848–1902*. Albuquerque: University of New Mexico Press.

———. 2002. *Water and American Government: The Reclamation Bureau, National Water Policy, and the West, 1902–1935*. Berkeley: University of California Press.

Platt, Harold L. 1986. "Samuel Insull and the Electric City." *Chicago History* 15 (1): 20–35.

———. 1991. *The Electric City: Energy and the Growth of the Chicago Area, 1880–1930*. Chicago: University of Chicago Press.

Plum, Lester V. 1938. "A Critique of the Federal Power Act." *Journal of Land & Public Utility Economics* 14:147–61.

Porter, Russell B. 1938a. "Press Lilienthal on TVA Cost Fixing." *New York Times*, July 24: 3.

———. 1938b. "TVA Power Lifts Farm Standards." *New York Times*, April 25: 21.

"Power Men Urged to Fight to Finish." 1935. *New York Times*, June 5: 7.

Priest, George L. 1993. "The Origins of Utility Regulation and the 'Theories of Regulation' Debate." *Journal of Law & Economics* 36 (1), part 2: 289–323.

Prins, Tony. 1991. "Niagara: Falling for Power." *IEE Review* 37 (10): 339–42.

Pritchett, C. Herman. 1942. "Administration of Federal Power Projects." *Journal of Land & Public Utility Economics* 18 (4): 379–90.

"Project Applications Received from 46 States." 1935. *Rural Electrification News* 1 (1): 9–10.

Proquest. 2015. "Proquest Historical Newspapers." Accessed July 16, 2015. http://www.proquest.com/products-services/pq-hist-news.html.

"Protests against Rayburn-Wheeler Bill Rise to Extraordinary Proportions." 1935. *Edison Electric Institute Bulletin* 3 (4): 105–7.

"P.S.C. Rulings Advance Wisconsin Rural Electrification." 1936. *Rural Electrification News* 2 (1): 19–22.

Public Utility Holding Companies. Part 3. 1935. Hearings before House Committee on Interstate and Foreign Commerce. April 5, 9, 10–12, 15, 74th Congress, 1st session.

Public Utility Holding Company Act of 1935. 1935. Hearings before Senate Committee on Interstate Commerce. April 16 to 29, 74th Congress, 1st session.

Purcell, Aaron D. 2002. "Struggle within, Struggle Without: The TEPCO Case and the Tennessee Valley Authority, 1936–1939." *Tennessee Historical Quarterly* 61 (3): 194–210.

"PWA-TVA 'Coercion' Charged by Utility." 1936. *New York Times*, October 14: 39.

"Question of Municipal Franchises, The." 1903. *Wall Street Journal*, March 16: 6.

"Question of Review, The." 1906. *Wall Street Journal*, February 24: 1.

Ramsay, M. L. 1937. *Pyramids of Power: The Story of Roosevelt, Insull, and the Utility Wars*. Indianapolis: Bobbs-Merrill.

Ramsey, F. P. 1927. "A Contribution to the Theory of Taxation." *Economic Journal* 37 (145): 47–61.

"Rate Clause Blocks Projects in the Northwest." 1936. *Rural Electrification News* 2 (1): 4, 31.

Raushenbush, Stephen. 1928. *High Power Propaganda*. New York: New Republic, Inc.

———. 1931. *The Power Fight*. New York: New Republic, Inc.

"REA Activities in Virginia Suspended." 1936. *Rural Electrification News* 1 (9): 7–8.

"REA Victory in Alabama Case Aids Fight against 'Spite Lines' in Other States." 1937. *Rural Electrification News* 3 (3): 3–4, 30.

"Reductions in Insurance Rates Obtained for REA-Financed Cooperatives." 1938. *Rural Electrification News* 3 (10): 7–8.

Reed, Hudson W. 1936. "Rural Electrification in the United States." In *Transactions of the Third World Power Conference*, 725–65. Washington, DC: Third World Power Conference.

"Report of Committee on Legislative Policy." 1904. In *National Electric Light Association, Twenty-Seventh Convention, Volume 1*, 89. National Electric Light Association.

Report of the Giant Power Survey Board to the General Assembly of the Commonwealth of Pennsylvania. 1925. Harrisburg, PA: Telegraph Printing Co.

"Report of the National Conservation Commission." 1909. S. Doc 676, 60th Congress, 2nd session.

"Report of the Rate Research Committee." 1914. In *National Electric Light Association, Thirty-Seventh Convention, Commercial Sessions*, 60–116. National Electric Light Association.

"Report of the Rural Lines Committee." 1922. In *Proceedings of the Forty-Fifth Convention of the National Electric Light Association, Volume 1*, 43–130. New York: National Electric Light Association.

"Report of the Superpower Survey Committee." 1921. In *Proceedings of the Forty-Fourth Convention of the National Electric Light Association, Volume 1*, 58–59. New York: National Electric Light Association.

Richardson, Lemont Kingsford. 1961. *Wisconsin REA: The Struggle to Extend Electricity to Rural Wisconsin, 1935–1955*. Madison: University of Wisconsin Experiment Station, College of Agriculture.

Ronayne, John H. 1946. "History of Farm Electrification." In *Farm Electrification Manual*, section 2: 1–3. New York: Edison Electric Institute.

Roosevelt, Franklin D. 1933. "A Request for Legislation to Create a Tennessee Valley Authority—a Corporation Clothed with the Power of Government but Possessed of the Flexibility and Initiative of a Private Enterprise," H. R. Doc 15, 73rd Congress, 1st session.

———. 1935. "Executive Order 7037 Establishing the Rural Electrification Administration." Accessed July 28, 2015. http://www.presidency.ucsb.edu/ws/?pid=15057.

————. 1937. "Fireside Chat 10: On New Legislation (October 12)." Accessed August 6, 2012, http://millercenter.org/president/speeches/detail/3311.

Roosevelt, Franklin D., and Samuel I. Rosenman. 1938. *The Public Papers and Addresses of Franklin D. Roosevelt—the Genesis of the New Deal: 1928–1932*. New York: Random House.

Rural Electrification. 1936. Hearings before House Committee on Interstate and Foreign Commerce. 74th Congress, 2nd session.

Rural Electrification Administration. 1964. *1963 Annual Statistical Report*. Washington, DC: US Government Printing Office.

S. 1796 to Amend the Tennessee Valley Authority Act of 1933. 1939. Hearings before House Committee on Military Affairs. 76th Congress, 1st session.

"Samuel Insull." 1938. *New York Times*, July 18: 12.

"Says Stock Slump Is Only Temporary." 1929. *New York Times*, October 24: 2.

Schiffer, Michael B. 2003. *Draw the Lightning Down: Benjamin Franklin and Electrical Technology in the Age of Enlightenment*. Berkeley: University of California Press.

Schlesinger, Arthur M., Jr. 1959. *The Coming of the New Deal*. Boston: Houghton Mifflin.

Schroeder, Fred E. H. 1986. "More 'Small Things Forgotten': Domestic Electrical Plugs and Receptacles, 1881–1931." *Technology and Culture* 27 (3): 525–43.

Scott, Ernest K. 1897. *The Local Distribution of Electric Power in Workshops, &C.* London: Biggs and Co. Accessed September 10, 2013. https://play.google.com/books/reader?id=928OAAAAYAAJ&printsec=frontcover&output=reader&hl=en&pg=GBS.PA9.

Sidak, J. Gregory, and Daniel F. Spulber. 1997. *Deregulatory Takings and the Regulatory Contract: The Competitive Transformation of Network Industries in the United States*. Cambridge: Cambridge University Press.

Slattery, Harry, and Sherman Fabian Mittell. 1940. *Rural America Lights Up*. Washington, DC: National Home Library Foundation.

Smalley, Harrison Standish. 1906. "Railroad Rate Control." *Publications of the American Economic Association* 7 (2): 327–473.

Spence, Clark C. 1962. "Early Uses of Electricity in American Agriculture." *Technology and Culture* 3 (2): 142–60.

Steffens, Lincoln. 1904. *The Shame of the Cities*. New York: McClure, Phillips & Co.

Sternegg, Bruno. 2014. "Singen Industrie." Accessed November 10. http://www.singen-hegau-archiv.ch/singen-industrie.html.

Stevens, Joseph E. 1988. *Hoover Dam: An American Adventure*. Norman: University of Oklahoma Press.

"Stone & Webster." 1930. *Fortune*, November: 91–98, 121.

Straffin, P. D., and J. P. Heaney. 1981. "Game Theory and the Tennessee Valley Authority." *International Journal of Game Theory* 10 (1): 35–43.

Sullivan, Brandon A. 2010. "Scandal and Reform: An Examination of Societal Responses to Major Financial and Corporate Crime." Master's thesis, Bowling Green State University.

Swain, George F. 1888. "Statistics of Water Power Employed in Manufacturing in the United States." *Publications of the American Statistical Association* 1 (1): 5–44.

Sweall, Stephan A., ed. 1913. "Question Box." *National Electric Light Association Bulletin* (9): 607–48.

Taft, Dub, and Sam Heys. 2011. *Big Bets: Decisions and Leaders That Shaped Southern Company*. Atlanta: Southern Company.

Talbert, Roy, Jr. 1987. *FDR's Utopian: Arthur Morgan of the TVA*. Jackson: University Press of Mississippi.

Taussig, F. W. 1926. *Principles of Economics*. 3rd ed. New York: Macmillan.

Taylor, Arthur R. 1962. "Losses to the Public in the Insull Collapse: 1932–1946." *Business History Review* 36 (2): 188–204.

Tennessee Valley Authority. 1935. Hearings before House Committee on Military Affairs. March 28-April 10, 74th Congress, 1st session.

Tennessee Valley Authority. 1934. "Annual Report of the Tennessee Valley Authority for the Fiscal Year Ended June 30, 1934."

———. 1935. *Annual Report of the Tennessee Valley Authority for the Fiscal Year Ended June 30, 1934*. Washington, DC: U.S. Government Printing Office.

———. 1936. *Annual Report of the Tennessee Valley Authority for the Fiscal Year Ended June 30, 1936*. Washington, DC: U.S. Government Printing Office.

———. 1939. "Annual Report of the Tennessee Valley Authority for the Fiscal Year Ended June 30, 1938."

———. 1940. "Annual Report of the Tennessee Valley Authority for the Fiscal Year Ended June 30, 1940."

———. 1941. "Annual Report of the Tennessee Valley Authority for the Fiscal Year Ended June 30, 1941."

———. 1956. *Rate Reductions by the Distributors of TVA Power*. Chattanooga.: Tennessee Valley Authority, Office of Power.

Tennessee Valley Authority, Vol. 2. 1935. Hearings before House Committee on Military Affairs. May 20–23, 1935, 74th Congress, 1st session.

Terral, Rufus. 1936. "See Pool Solving TVA Controversies." *New York Times*, October 4: E7.

"Text of Governor Roosevelt's Speech at Commonwealth Club, San Francisco." 1932. *New York Times*, September 24: 6.

"Texts of Roosevelt Speeches Urging Public Operation of Power Plants." 1934. *New York Times*, November 19: 3.

Thompson, Carl D. 1932. *Confessions of the Power Trust*. New York: E. P. Dutton & Co.

Tomain, Joseph P. 2002. "The Past and Future of Electricity Regulation." *Environmental Law* 32 (2). Accessed May 2, 2015. http://go.galegroup.com/ps/i.do?id

=GALE%7CA87710972&v=2.1&u=gree35277&it=r&p=ITOF&sw=w&asid=f733db39388f47dc824b2b2fac35da93.

"Transactions on the New York Stock Exchange." 1933. *New York Times*, September 16: 20.

Troesken, Werner. 1997. "The Sources of Public Ownership: Historical Evidence from the Gas Industry." *Journal of Law, Economics, and Organization* 13 (1): 1–25.

———. 2006. "Regime Change and Corruption: A History of Public Utility Regulation." In *Corruption and Reform: Lessons from America's Economic History*, edited by Edward L. Glaeser and Claudia Goldin, 259–81. Chicago: University of Chicago Press.

Troesken, Werner, and Rick Geddes. 2003. "Municipalizing American Waterworks, 1897–1915." *Journal of Law, Economics, and Organization* 19 (2): 373–400.

Twentieth Century Fund, The. 1948. *Electric Power and Government Policy: A Survey of the Relations between the Government and the Electric Power Industry*. New York: The Twentieth Century Fund.

Ulmer, Melville J. 1960. *Capital in Transportation, Communications, and Public Utilities: Its Formation and Financing*. Princeton: Princeton University Press.

US Army Chief of Engineers, and Secretary of the US Federal Power Commission. 1926. "A Letter Showing All Navigable Streams Upon Which Power Development Appears to Be Feasible and the Estimate of Cost of Examinations of the Same." H. R. Doc 308, 69th Congress, 1st session.

US Bureau of Corporations. 1912. *Water-Power Development in the United States*. Washington, DC: Government Printing Office.

US Bureau of the Census. 1905. *Central Electric Light and Power Stations, 1902*. Washington, DC: US Government Printing Office (reprinted 2006, New York: Ross Pub.).

———. 1910. *Central Electric Light and Power Stations, 1907*. Washington, DC: US Government Printing Office (reprinted 2006, New York: Ross Pub.).

———. 1915. *Central Electric Light and Power Stations, 1912*. Washington, DC: US Government Printing Office (reprinted 2006, New York: Ross Pub.).

———. 1920. *Central Electric Light and Power Stations, 1917*. Washington, DC: US Government Printing Office (reprinted 2006, New York: Ross Pub.).

———. 1924. *Biennial Census of Manufactures, 1921*. Washington, DC: US Government Printing Office.

———. 1930. *Central Electric Light and Power Stations, 1927*. Washington, DC: US Government Printing Office (reprinted 2006, New York: Ross Pub.).

———. 1932. *Fifteenth Census of the United States (1930). Agriculture, Vol. IV: General Reports*. Washington, DC: US Government Printing Office.

———. 1933. *Fifteenth Census of the United States. Manufactures: 1929*. Washington, DC: US Government Printing Office.

———. 1934. *Central Electric Light and Power Stations, 1932*. Washington, DC: US Government Printing Office (reprinted 2006, New York: Ross Pub.).

————. 1939. *Electric Light and Power Industry, 1937.* Washington, DC: US Government Printing Office (reprinted 2006, New York: Ross Pub.).

————. 1940. *Sixteenth Census of the United States (1940), Manufactures, 1939.* Washington, DC: US Government Printing Office.

————. 1966. *1963 Census of Manufactures.* Washington, DC: US Government Printing Office.

————. 1975. *Historical Statistics of the United States, Colonial Times to 1970, Vol. II.* Washington, DC: US Government Printing Office.

US Commissioner of Labor. 1900. *Fourteenth Annual Report of the Commissioner of Labor, 1899: Water, Gas, and Electric-Light Plants under Private and Municipal Ownership.* Washington, DC: US Government Printing Office.

US Department of Agriculture. 1916. "Electric Power Development in United States and Concentration in Ownership and Control, Part 1," S. Doc 316, 64th Congress, 1st session.

US Energy Information Administration. 2014a. "Electricity Data Browser." Accessed August 3. http://www.eia.gov/electricity/monthly/.

————. 2014b. "Electricity Data." Accessed June 27. http://www.eia.gov/electricity/data.cfm#generation.

US Federal Power Commission. 1940. *Typical Electric Bills.* Washington, DC: US Federal Power Commission.

US Federal Trade Commission. 1927. "Control of Power Companies," S. Doc. 213, 69th Congress, 2nd session.

————. 1928a. "Electric Power Industry: Supply of Electrical Equipment and Competitive Conditions," S. Doc 46, 70th Congress, 1st session.

————. 1928b. "Supply of Electrical Equipment and Competitive Conditions," S. Doc. 46, 70th Congress, 1st session.

————. 1930. "Utility Corporations, No. 22, American Gas and Electric Co., Electric Bond and Share Co.," S. Doc 92, Part 22, 70th Congress, 1st session.

————. 1934. "Summary Report: Efforts by Associations and Agencies of Electric and Gas Utilities to Influence Public Opinion," S. Doc. 92, Part 71-A, 70th Congress, 1st session.

————. 1935a. "Summary Report on Economic, Financial, and Corporate Phases of Holding and Operating Companies of Electric and Gas Utilities," S. Doc. 92, Part 72-A, 70th Congress, 1st session.

————. 1935b. "Summary Report: Survey of State Laws and Regulations, Present Extent of Federal Regulation and the Need of Federal Legislation, Conclusions and Recommendations, and Legal Studies in Support Thereof," S. Doc. 92, Part 73-A, 70th Congress, 1st session.

————. 1935c. "Utility Corporations No. 76 Duke Power Co. System," S. Doc. 92, Part 76, 70th Congress, 1st session.

US House Committee on Interstate and Foreign Commerce. 1934a. "Relation of Holding Companies to Operating Companies in Power and Gas Affecting Control—Part 2," H. Rep. 827, 73rd Congress, 2nd session.

———. 1934b. "Relation of Holding Companies to Operating Companies in Power and Gas Affecting Control—Part 3," H. Rep 827, Part 3.

US House Committee on Rivers and Harbors. 1930. "Report of the Chief Engineer on the Tennessee River and Tributaries, Part 1," H. R. Doc. 328, 71st Congress, 1 session.

US Joint Committee on the Investigation of the Tennessee Valley Authority. 1939. "Report of the Joint Committee on the Investigation of the Tennessee Valley Authority," S. Doc 56, 76th Congress, 1st session.

US National Park Service. 2012. "Roosevelt Dam Withdrawal of National Historic Landmark Designation." Accessed June 26. http://www.nps.gov/nhl/DOE_de designations/Roosevelt.htm.

US Rural Electrification Administration. 1938a. *1937 Report of Rural Electrification Administration*. Washington, DC: US Government Printing Office.

———. 1938b. *Rural Electrification on the March*. Washington, DC: Rural Electrification Administration.

———. 1939. *1938 Report of Rural Electrification Administration*. Washington, DC: US Government Printing Office.

———. 1940. *1939 Report of Rural Electrification Administration*. Washington, DC: US Government Printing Office.

———. 1941. *1940 Report of the Administrator of the Rural Electrification Administration*. Washington, DC: US Government Printing Office.

———. 1966. *Rural Lines, USA: The Story of Cooperative Rural Electrification*. Washington, DC: US Dept. of Agriculture.

———. 1968. *1967 Report of the Administrator Rural Electrification Administration*. Washington, DC: US Dept. of Agriculture.

———. 1985. "A Brief History of the Rural Electric and Telephone Programs."

US Securities and Exchange Commission. 1952. *Eighteenth Annual Report*. Washington, DC: US Government Printing Office.

———. 1959. *25th Annual Report of the Securities and Exchange Commission*. Washington, DC: US Government Printing Office.

US Senate Committee on Banking and Currency. 1934. "Stock Market Practices (Pecora Commission)," S. Rep. 1455, 73rd Congress, 2nd session.

"Utilities Get Call to 'Fight for Life'." 1935. *New York Times*, June 6: 31.

"Utility Findings Are Called Fraud." 1935. *New York Times* June 4: 31.

"Utility Issues Hit by Federal Rates." 1933. *New York Times*, September 16: 19.

"Utility Men Hail Birmingham Vote." 1933. *New York Times*, October 11: 37.

Van Derzee, G. W. 1928. "Rural Electric Service." In *1928 AEIC Minutes*, 1073–151. New York: Association of Edison Illuminating Companies.

Voeltz, Herman C. 1962. "Genesis and Development of a Regional Power Agency in the Pacific Northwest, 1933–43." *Pacific Northwest Quarterly* 53 (2): 65–76.

Wagner, Herbert A. 1900. "The Use of Alternating Current for the Extension of Central Station Supply and for General Distribution (with Discussion)." In

1900 AEIC Minutes, 112–33. New York: Association of Edison Illuminating Companies.

Wahl, Richard W. 1989. *Markets for Federal Water: Subsidies, Property Rights, and the Bureau of Reclamation*. Washington, DC: Resources for the Future.

Wald, Matthew L. 2013. "Ideas to Bolster Power Grid Run Up against 500 Owners of Different Minds." *New York Times*, July 15: A11.

Walker, H. B. 1928. "Research in Mechanical Farm Equipment." Washington, DC: US Department of Agriculture.

Walthall, E. B., E. P. Edwards, F. E. Cronise, E. L. Montgomery, and S. M. Kennedy. 1912. "Report of the Committee on Electricity in Rural Districts." Paper presented at the 35th Convention of the National Electric Light Association, New York. Vol. II, *Commercial Sessions*, 262–329.

Waterman, Merwin H. 1936. *Public Utility Financing, 1930–35*. Ann Arbor: University of Michigan School of Business Administration Bureau of Business Research.

Wells, Phillip Patterson. 1927. *Report of the Giant Power Board to the Governor of Pennsylvania*. Harrisburg, PA.

"Welsbach Light, The." 1900. *Science* 12 (312): 951–56.

"Wendell Lewis Willkie." 1937. *Fortune* 15: 89–90, 202–9.

Wengert, Norman. 1952. "Antecedents of TVA: The Legislative History of Muscle Shoals." *Agricultural History* 26 (4): 141–47.

What About Rural Electrification? Proceedings of the Farm Electrical Conference. 1926. Chicago, May 12–13.

"What May Happen." 1905. *New York Times*, May 15: 8.

White, Eugene N. 1990. "The Stock Market Boom and Crash of 1929 Revisited." *Journal of Economic Perspectives* 4 (2): 67–83.

White, Leonard D. 1921. "The Origin of Utility Commissions in Massachusetts." *Journal of Political Economy* 29 (3): 177–97.

Wigmore, Barrie A. 1985. *The Crash and Its Aftermath: A History of Securities Markets in the United States, 1929–1933*. Westport, CT: Greenwood Press.

Wilbur, Lyman, and Northcutt Ely. 1933. *The Hoover Dam Power and Water Contracts and Related Data*. Washington, DC: US Government Printing Office.

Wilcox, Delos F. 1910. *Municipal Franchises*. Vol. 1. New York: McGraw-Hill.

"Will Confer with Utilities." 1936. *New York Times*, September 12: 3–3.

Williamson, Oliver E. 1979. "Transaction-Cost Economics: The Governance of Contractual Relations." *Journal of Law and Economics* 22 (2): 233–61.

———. 1985. *The Economic Institutions of Capitalism: Firms, Markets, Relational Contracting*. New York: Free Press.

———. 1988. "The Logic of Economic Organization." *Journal of Law, Economics, & Organization* 4 (1): 65–93.

Willis, Hugh. 1927. "The Rate Base for Rate Regulation." *Indiana Law Journal* 3 (3): 225–32.

Wilson, Andrew. 2002. "Machines, Power and the Ancient Economy." *Journal of Roman Studies* 92:1–32.

Wolfram, Catherine D. 1999. "Electricity Markets: Should the Rest of the World Adopt the United Kingdom's Reforms?" *Regulation* 22 (4).

Wolfe, Audra J. 2000. "'How Not to Electrocute the Farmer': Assessing Attitudes Towards Electrification on American Farms, 1920–1940." *Agricultural History* 74 (2): 515–29.

Wright, Arthur. 1896. "Cost of Electricity Supply." In *Municipal Electrical Association Proceedings*, 44–67. Reprinted 1915 in *The Development of Scientific Rates for Electricity Supply*, 31–51, Detroit: Edison Illuminating Company.

———. 1897. "Profitable Extensions of Electricity Supply Stations." Paper presented at the 20th Convention of the National Electric Light Association, Niagara Falls, NY, June 8–10, 159–89.

Wunderlin, Clarence E. 1992. *Visions of a New Industrial Order: Social Science and Labor Theory in America's Progressive Era*. New York: Columbia University Press.

Young, Harold H. 1965. *Forty Years of Public Utility Finance*. Charlottesville: University Press of Virginia.

Zinder, Hanina. 1929. "Problems of Rural Electric Service: Rural Rates and the Financing of Rural Line Extensions." *Journal of Land & Public Utility Economics* 5 (1): 79–89.

Index